EVOLUTION OF PHYSICAL LAWS

EVOLUTION OF PHYSICAL LAWS

A theory of General Physics by
Satdev Sharma
*(Former Head of Chemistry Department
Layallpur Khalsa College, Jalandhar)*

Copyright © S.D. Sharma, 2006

Dedicated to

My Wife and My Parents

" Of all that doth exist what is the final cause?

Thus do the wise ones ask, who do the Vedas seek –

Brahman, or something else? Whence all that's here once was?

Where doth it end at last? From where rose joy and grief?"

(Shvetaashvatara Upanishad, Chapter 1 Verse I)

PREFACE

The humble attempt is made in the present short thesis to give mathematical and model form to the ideas about creation expressed in the seed form in the ancient scriptures. **Evolution of physical laws** is the discipline code for the matter to follow to keep the chaos under check. The evolution of physical laws is necessary consequence of the evolution of matter. The model of the basic units of mass are Akashon (etheron) and that of energy is Prakashon (photon) are governed by very simple mathematical relations and all the laws governing the Universe can be easily dug out by the proper handling of the model and the associated simple mathematical relations without using mind boggling higher mathematics. Etheron is got from ether and Prakashon is linked to *Prakasha*, which means light in *Sanskrit* language.

The present treatment is continuous without any chapters, because there is single source of all the theories and that source as already mentioned is the basic Akashon or Prakashon. All the physical laws are packed in the single fruit of knowledge and the proper lancet sharpened by the human intellect is needed for dissection to unravel the secrets. It is not necessary that the abstract higher mathematics is the only proper instrument. Even the simple model can explain the things in such a way that even a layman can understand the complex theories. Inspired by the philosophical and meta-physical generalization, the new treatment has been proposed to understand the theories of modern physics. The present treatment gives concrete form to the abstract ideas.

I acknowledge the valuable suggestions of my wife Naresh Sharma regarding this book. I am also thankful to my son Nikhilesh and daughter-in-law Taruna for their help in proof reading. My son Somesh and daughter-in-law Rutool deserve special thanks for the help rendered in typing the manuscript and sketching the diagrams.

Jalandhar **Satdev Sharma**

Feedback & Reception of the theory

(The Tribune)

Physics through oriental eyes

MINNA ZUTSHI
TRIBUNE NEWS SERVICE

Seeing old things in new light is the essence of creativity. And ancient texts can very well be seen in a refreshingly fresh context if the perspective is given a slight shift, feels Mr S.D. Sharma, Head of Chemistry Department at Lyallpur Khalsa College here. He has penned a book on evolution of physical laws.

Interestingly, he has given a different treatment to this very scientific subject. He has taken Akashon (mass packet) and Prakashon (photon) as the realistic models of fundamental units of mass and energy. He traces the evolution of physical laws, right from transcendental First Cause to creative energy of maya. "I have taken a cue from our ancient texts. Starting with The First Principle as mentioned in the ancient scriptures, I have gravitated down to the gross creation associated with physical laws that govern the universe," he explains.

S.D. Sharma

Mr. Sharma, whose book Evolution of Physical Laws (A Theory of General Physics) is his first foray into scientific writing, says that the oriental sages expressed their views about the nature of creation in seed form.

"There is no further useful analytical elaboration of the thought. I have made an attempt to associate this thought with mathematical symbols to make it fit for quantitative studies," he explains, adding that the concepts of three gunas and maya can be interpreted in terms of the fundamental qualities of nature, force, inertia etc.

The author claims that he has made an attempt to introduce concepts from varied fields in a unified manner. He has done away with the usual practice of dividing a book into chapters. Theories of relativity and quantum mechanics find a rather interesting interpretation in his book. "I have tried to dissolve the dichotomy between science and religion," he adds.

What are new Concepts in the Book?

⇨ Subtle Akasha or ether is raw material for the creation of illusory gross phenomenal universe. Entirely new concept of ether or Akasha is introduced. Ether or Akasha is the alpha and omega of creation.

⇨ All things are unified because there are no boundaries of objects. The boundaries are only apparent.

⇨ Units of mass (Akashon) and energy (Prakashon) are simple arrangements containing Akasha or ether. Dimensions of Akashon and Prakashon are limitless.

⇨ The relations for mass and energy are not empirical in nature but are logical.

⇨ Each Akashon has got limiting velocity called terminal velocity.

⇨ Entirely new concept of field disregarding paradigm concept.

⇨ There is a limit to the maximum force.

⇨ The elementary particle with heaviest mass is introduced.

⇨ Heisenberg's uncertainty principle, dual nature and quantum mechanics adopting new approaches.

⇨ The equivalence of inertial and gravitational masses is not empirical but is logical.

⇨ The special and the general theory of relativity are discussed in such a manner that even a layman can understand the complex theories.

⇨ The general theory of relativity has been discussed retaining the three-dimensional real space.

⇨ The energy problems of quantum mechanics have been discussed by using classical mechanics and very simple mathematics.

⇨ The concept of electromagneton and mattenergon has been introduced.

⇨ Simple experiment to prove existence of Akasha or ether.

List of Topics

S.no.	Topic	Page
1.	Metamorphosis of Unmanifested to Manifested State	11
2.	The Existence of Ether or Akasha (Space)	16
3.	Use of Models to Explain the Theories of Modern Physics	25
4.	The Birth of Field, Matter (Akashon) and Energy (Prakashon)	27
5.	Derivation of Mother of All Equations	38
6.	The Force and Energy	46
7.	No Observation of Singularity	56
8.	The Force and Inertia	57
9.	Creation of Temporary Region of Perturbation	64
10.	Determination of the Constants Associated with the Equation of the Principle of Inverse Variation	67
11.	Derivation of Force Relation	71
12.	Measurement of Force and Derivation of Force Relation	74
13.	The Cosmic Terminal Velocity of Elementary Particle	91
14.	Benchmark Velocity and Change in Velocity of Light in Vacuum	96
15.	Equivalence of Mass and Energy	101
16.	Size of Photon	103
17.	Properties of Akashons and Prakashons	106
18.	The Big Bang and Black Holes	112
19.	Dark Matter and Dark Energy	117
20.	The Black Hole	119
21.	The True Akashon	120
22.	Special Theory of Relativity	124

S.no.	Topic	Page
23.	Relative Measurement of Length and Time	127
24.	Measurement of Time	130
25.	General Expression for the Measurement of Length and Time in the two Frames Moving with Relative Velocity	137
26.	Concept of Field	141
27.	The Manifestation of Gravitational Field	143
28.	Derivation of Newton's Gravitational Force Relation	147
29.	Equivalence of Inertial Mass and Gravitational Mass	155
30.	Special Case of Interaction	156
31.	The Proof that the Region of Perturbation Contains Oscillating Shells	160
32.	The Gravitational Constant and Super Heavy Akashon (N)	161
33.	Super Heavy Akashon or Etheron (or N Akashon)	165
34.	General Theory of Relativity	168
35.	Calculation of Mass of Elementary Particles	193
36.	Quarks	200
37.	The Fractional Charge on the Quarks	203
38.	The Stability of Elementary Particles	204
39.	General Gravitational Interaction of two Akashons	211
40.	The Value of k_t	212
41.	Guidelines to Observe Levitation or to Overcome Force of Gravity	213
42.	Spooky Action at Distance in Plants	214
43.	The Charge and Electric Field	220
44.	Derivation of Coulomb's Law of Force Between two Charges	224
45.	Calculation of Elementary Charge and Types of Charge	229

S.no.	Topic	Page
46.	Metamorphosis and Charge Packet	234
47.	Self Electrical Potential Energy of Charged Akashon and Electromagneton	235
48.	The Acceleration of the Charged Akashon	240
49.	Nature of Photon and Photon Energy	247
50.	The Birth of Photon and Energy of Photon	259
51.	Association of Electric and Magnetic Fields with Photon	264
52.	The Self Gravitational Potential Energy of Pure Mass of Akashon and Hybrid Akashon	269
53.	The Unification of Electromagnetic Field and Gravitational Field	272
54.	The Proofs in Favour of the New Model of Photon	277
55.	Interaction of Particles	279
56.	Casimir Effect	284
57.	Magnetic Field	290
58.	Dual nature of Matter in State of Motion	301
59.	De-Broglie wave Equation	309
60.	Michelson-Morley Experiment	312
61.	Dual Nature of Photon	319
62.	Explaining the Dual Nature of Prakashon (Photon) and the Moving Akashon	323
63.	Double Slit Diffraction Pattern of Electromagnetic Radiation and Moving Akashon	324
64.	Heisenberg Uncertainty Principle	330
65.	Mattenergon Quantum Mechanics or Wave Mechanics	353
66.	Particle in a Box	363
68.	Simple Harmonic Oscillator	366
69.	The Akashon (Etheron) Moving in the Circular Path and Rigid Rotator	371
70.	Hydrogen Like Atom	385

S.no.	Topic	Page
71.	The Relation Between the Orbital Quantum Number 'l' and Principle 'n'	401
72.	Spin Angular Momentum of the Akashon or the Mass Packet	404
73.	Spin Magnetic Moment of the Charged Mass Packet	412
74.	The Fusion of two Akashons	415
75.	Unio (Onion) – Model of Nucleus	417
76.	Radioactivity	425
77.	Fission of the Nucleus	429
78.	Transmutation of Elements	431
79.	The Unification of Forces	432
80.	The Weak Force	439
81.	The Nature of Strong Force	442
82.	The Human Conscious	447
	Glossary	

Abraham Lincoln once said that his politics was like that of old women's dance, so is my methodology to understand the theories of higher physics by associating the ideas expressed in the Hindu scriptures in the seed form with very simple mathematics.

METAMORPHOSIS OF UNMANIFESTED TO MANIFESTED STATE:

In the Upanishads, *Hindu* scriptures, Brahma, the supreme source of creation, is believed to be self existent, all pervading, eternal from which all beings emanate and to which all return. It is uncreated, unborn and without beginning. The transcendental Brahma is devoid of qualifying attributes. The Brahma is independent of causation. The law of cause and effect operates only when the Brahma is manifested as the phenomenal world in association with "Maya", the prime cause of creation. The phenomenal world is experienced as effect of prime cause of creation "The Maya". The absolute emptiness cannot give birth to something gross without involvement of the creative agency the Maya. The Brahma is the substratum of the manifested universe. The Upanishads describe that as the threads come out of a spider; as waves move on the surface of the ocean; as sparks scatter away from the burning fire; so also has the Universe come from Brahma. The Brahma is present in the creation not in parts but as undivided whole. The illusion of plurality is the effect of Maya. The non-dual Brahma is manifested as Universe with the attributes of plurality in association with Maya. Vedantist philosophers described Maya a power of Brahma. After the creation of the phenomenal world, the effect of Maya dominates the stage, while the absolute one is hidden behind the curtains of illusions. The three attributes or *gunas* of Maya are Sattva, Tamas and

Rajas. The three mundane directors manage the illusory stage show named as creation. Sattva guna stands for force, Tamas stands for inertia while the Rajas represent activity or energy. Acutally the two gunas, Tamas and Satva are the primary gunas while the third guna Rajas is the derived guna because it is result of the combination of Sattva and Tamas. It is the force (Sattva), which imparts motion (Rajas) to the stone at rest (Tamas). There is no creation so long as the gunas are dormant or in the seed form. The Rigveda in the following statement virtually describes the so-called Big Bang theory of creation. "The darkness was there at first by darkness covered. The world was ocean without destination. But a poignant germ lay hidden in the shell. The one engendered by force of heat." There has been intense curiosity about the source and cause of creation in the minds of sages of yore. The roots of the modern science arein the *vedic* philosophy, therefore, getting cues from there, the mystery of creation can be easily unraveled.

The ideas in the seed form related to creation expressed in the abstract synthetic philosophical language of the Vedas and Upanishads can be elaborated so that the seed form of thought sprouts to ever expanding tree of knowledge which helps in understanding the phenomenal world in terms of the mundane analytical scientific terminology. The phenomenal world is governed by the chain of cause and effect initiated by the primary cause. The primary cause of creation is very simple when divested of the abstract philosophical implications. The primary cause of creation can be explained in simple language as given below.

The motion or perturbation is the primary cause of creation of the physical Universe along with the natural laws, which keep the chaos under control because the perturbation or motion without restraints leads to chaos.

Subtle (Akasha) $\xrightarrow{\text{Perturbation}}$ Gross creation

The preferred motion is repetitive in natue like circular or oscillatory motion. The physical laws give semblance of order to the chaotic physical creation when viewed on the macroscopic scale. What is experienced as the manifested creation is **the effect of the primary cause, which is the perturbation or motion**. The truth is always simple. The above simple observations can be applied to some observations in the created world we live in. A lay man riding on a bike attributes the balanced state of the bike to motion because a stationary bike is unbalanced. The production of heat or charge on rubbing together two things is considered as the effect of motion. So many examples can be sited at the gross level related to the effect of motion. Magnetic field is the effect of the motion of the charge. Without motion there is no change and with change, time is born. Kinetic energy is the effect of active motion while potential energy is the effect of passive motion because potential energy is the effect of change of position and change of position requires motion. As already stated, the fundamental creative agency Maya is motion or perturbation when interrupted in gross terms. The production of musical notes from a wire is a two stage process. The first stage involves the tightening of the wire with force (Sattva process) while the second stage requires the plucking or hammering of the wire (Tamas state changes to Rajas state). The two processes requie foce. The above two processes should be performed one after the other in the order given above. It shall be seen later on that the manifested creation like the production of musical notes from strings is also a two stage process, and during the two stage process of creation, the primal sound of creation 'OM' reverberated all around. The myriads of sweet musical notes can be produced by altering the mode of tightening of the wire in the first stage and adopting the different ways of plucking the wire in the second stage. The same thing is applicable to the manifested creation which is the cosmic symphony directed by the primary cause, Maya. The derived guna Rajas (perturbation) of the creative agency Maya dominates the manifested creation around and that is why the stress is laid on Shanti (tranquillity) in the vedic hymns to keep the things under control.

The absolute stillness is the fundamental attribute of the transcendental Brahms. In association with motion or perturbation the non-dual Brahma creates phenomenal world with attributes of plurality. The phenomenal world can be interpreted as interplay of three gunas of Maya, Rajas (energy), Tamas (matter) and Sattva (field) or actually two gunas Tamas and Sattva because as already mentioned, Rajas is a derived guna. It shall be seen later on that the matter, energy and the field are the effects of the fundamental cause of creation, the perturbation (Maya). **There is no creation without perturbation and that is why the creative agency Maya is said to be the cause of illusion.** The same wires or the same flute controlled by deft fingers of a musician can produce numerous musical notes.

The *Upanishads* describes the cosmos as universal motion. **Maya (motion) is the primary cause of the birth of the gross out of subtle.** The cause of manifestation of Brahma (absolute stillness) or the birth of mutable gross matter is the cosmic dance of Shiva (motion). The cosmic dance is the controlled motion but perturbation without any restraint can lead to chaos or destruction (Rudra form). The gross manifestation of Brahma is sustained by Vishnu, the preserver through the evolution of field which keeps the perturbation under control. As the divine Triune is one so also are the gross manifestations field, matter and energy unified because the gross manifestation is the effect of the same cause, motion on the same substratum.

THE EXISTENCE OF ETHER OR AKASHA (SPACE):

No sophisticated experiments like ridiculous Michalson-Morley experiments are needed to detect the presence of ether because experiment was performed to detect that all state of ether which was prior to its manifestation or conversion into mass and energy.The unmanifested ether (akasha) was converted to mass units(akashons) and energy units (prakashons) at the time of creation, therefore,Michelson-Morley experiment was performed to detect that conventional or primordial state of ether(akasha) in which it was present prior to its manifestation as mass or energy and space is devoid of that conventional state therefore the experiment led to null results apart from invariance of light speed discussed later on .Prior to big bang all the unmanifested subtle ether (akasha) was concentered at nearly point volume which exploded to give birth to creation or gross form of ether(akasha) After the creation all of the all pervading ether(akasha) was changed to morphosed state or gross state as mass and energy. Gross form of akasha is named as morphosed state instead of metamorphosed state because change of gross akasha to subtle akasha is reversible while metamorphosis is irreversible hence the new word was coined. At present the space is devoid of the pristine state of akasha The influence of morphosed state concentrated as mass packet(akashon) or energy packet(prakashon) extends upto infinity with steep fall in density with increase in distance from the centres of packets as explained later on quantitavely The morphosed akasha(ether) having practically negligible density present in space after creation is manifested as field and thus participate in spooky action at distance as explained later on while quantifying the field The mechanism of conversion of ether(akasha) into mass – energy is explained later on At present the space is not filled with ether on which Michelson –Morley experiment was basedThe concept of ether(akasha)was initially proposed to explain wave propagation of light so how to explain wave nature if no ether is present in space The akasha (ether)required for exhibition of wave nature is carried by photon and is not supplied by space as was misconceived by Michelson-Morley prior to experiment that light waves travel in stationary ether. No doubt medium is required for wave nature but that medium the conventional ether is carried by photon as army tank carries caterpillar track. How it happens is discussed while discussing nature of photon. The novel concept of ether is used in the present text.The conventional all pervading ethe(akasha)was there prior to creation and the conventional state disappeared after conversion to matter and energy therefore transformed ether(akasha)is all; around in morphosed state and the presence of ether(akasha) in changed state as matter and energy can not be denied Moreover the exact and experimentally proved results derived by using classical mechanics based on novel concept of ether(akasha) is strong proof that everything is composed of ether (akasha) no experiment is needed to prove existence of that which is self evident The creation is simply gross form of subtle form of ether(akasha) or morphosed state of akasha The main stream scientists have followed circuitous paths to reach a goal but in my book Evolution of Physical Laws i have achieved the same results accurately by following the straight path To understand basic fundamentals of nature there is no need to discard commonsense as alleged by theoretical physicists of repute My theory is based on not that old childish concept and state of ether Michelson –Morley experiment failed to detect, my concept is novel and original. The existence of ether is self-evident because ether is the substratum of all tht is manifested. The creation just did not precipitate out of nothing. The change of gross to subtle and the reverse process goes on perpetually. The simple experiments on plants reveal the existence of ether. A *Hindu* story tells of a fish who asked of another fish, "I have always heard about the sea, but what is it? Where is it?" The other fish replied, "you live, move, and have your being in the sea. The sea is within you and without you, and you are made of sea, and you will end in sea. The sea surrounds you as your own being." In this story we can read ether instead of sea and thus the nature of ether is succinctly defined. Learn from the lowly fish the true nature of Akasha or ether. There is no doubt about the existence of ether because the source of all that is manifested is ether. The conventional concept of the ether is akin to stationary fluid, something different from the rest of creation, but actually it is not so. In the above story, sea is just extension of manifested creation or fish. Whence everything is composed of ether, there is no boundary line separating the manifested creation and ether or Akasha. Euphemism for the ether is Akasha which means space in *Sanskrit* language. In subsequent writings, the word ether can be replaced with space or *Akasha*. There has been intense curiosity about the creation of manifested world. Here I quote from the *Hindu* Scriptures,

> *" Of all that doth exist what is the final cause?*
> *Thus do the wise ones ask, who do the Vedas seek –*
> *Brahman, or something else? Whence all that's here once was?*
> *Where doth it end at last? From where rose joy and grief?"*

(Shvetaashvatara Upanishad, Chapter 1 Verse I)

No one has seen virtual particles and to me there existence is doubtful. The concept of virtual particles is retained because it helps in understanding some observed phenomenon. The same phenomenon can be easily explained by the concept of ether. The method adopted here is much easy and appeals to the common sense. The concept of ether can easily be explained by the field and particle interactions. I have explained nature of fields without involving the unrealistic virtual particles. What is the harm in using models of mass packet and energy packet based on ether or Akasha or you can name anything when theories of physics can be explained without using complex mathematics.Eye can see everything around what fails to see itself.

In another *Hindu* Scripture *'Chhandogaya Upnishada' Pravahana* says to *Silaka* that the space known as *Akasha* (Ether) is the origin, support, and end of all. There is no further useful analytical elaboration of the "seed thought". I have made an attempt to nurture the seed of knowledge by associating it with simple mathematical equations.During the nineteenth century it was commonly assumed by many that matter was embedded in ether, therefore when object moves it has to plough through the ether but my concept of ether on which my theory is based is entirely different. The ether or Akasha is the alpha and omega of creation. Actually ether (Akasha) is the progenitor of mass and energy. Mass and energy and force are the temporary manifestations of ether. How ether manifests as matter and energy will be shown in the following pages. Based on this one point and other points it will be shown later on that Michelson-Morley experiment is inadequate to detect the presence of ether. Einstein denied the existence of ether for only 11 years (1905-1916), thereafter he recognized that his attitude was too radical and even regretted that his works published before 1916 had so definitely and absolutely rejected the existence of ether. Henceforth the terms ether and Akasha will be synonymous. Michelson-Morley experiment failed to detect the presence of ether because the experimental set up was based on naive concept of ether according to which there was prior assumption that ether is stationary. Therefore, light waves move in the ether as waves move on the surface of pond water. It is so whenever we think of ripples we are conditioned to see waves on the surface of water. **This childish and naive concept of ether on which Michelson-Morleyexperiment was based led to null results.** There is nothing to be surprised at the null results. The perturbation of Akasha and anti Akasha shells results in the manifestation of gross matter and energy. Actually ether is raw material for the formation of matter and energy units, Akashon (mass packet) and Prakashon (Photon). Akashon is after the word *Akasha*, which means space in *Sanskrit* and Prakashon is after the word *Prakasha* which stands for light. The ether which is associated with the formation of Akasha and Prakasha is called bonded ether, it is part and parcel of units of mass and energy. The non-bonded ether is that ether shells, which are left over and it is not associated with units. While the Akashon and Prakashon are moving all the bonded ether also moves hence the bonded ether does not experience any obstruction because no ether is present in the conventional space.

The existence of anything cannot be denied if the experimental set up to detect presence is inadequate and the experiment is based on the false properties of the thing to be detected. To confirm the presence of manifested existence, the usual process is that of division of experimental setup into subject and object which are considered to be distinct but in case of ether, no such schism is possible because everything that is manifested is ether. No definite boundary can be assigned to manifested existence. The boundary tapers off steeply and extends upto infinity. Due to steep tapering off only the unrealistic boundary can be only assigned. Actually everything is boundless. The above logical observation points out that the existence of ether cannot be poved by adopting the usual way of subject and object labels. There is no distinction between object and subject therefore, the negative results of Michelson-Morley using arrangements of mirror are expected. The positive results would have been contrary to the true nature of ether.A blind man cannot deny the existence of world around him because there are other ways for him to feel the indirect presence of the creation. Starting with the logical concept of ether I have derived all the known and unknown results of nature. Thereby, proving that existence of ether is not false notion. Even some of the results have been derived, which are being assumed as such so far without any valid derivation. Starting with the concept of ether the internal structure of units of mass Akashon and units of energy Prakashon have been elucidated for the first time.

In the words of Einstein, "it would have been more correct if I had limited myself to my early publication to emphasizing only the non-existence of an ether velocity instead of arguing the total non-existence of the ether, for I can see that with word ether we say nothing else than that the space has to be viewed as carrier of physical qualities. In 1905 I was of the opinion that it was no longer allowed to speak about the ether in physics. This opinion, however, was too radical as we will see later when we discuss the general theory of relativity. It does remain allowed, as always, to introduce a medium filling all the space and to assume that the electromagnetic fields (and matter as well) are its states. Once again empty space appears as endowed with physical properties i.e. no longer physically empty as seemed to be the case according to the special relativity. There is an important argument in favour of hypothesis of the ether. To deny the existence of ether means, in the last analysis, denying all physical properties to empty space."

The concept of ether in the present text shows that ether or Akasha does not offer any impedance to the movement of material bodies. When the Akashon or Prakashon moves all the bonded shells associated with these move. The dimensions of Akashons and Prakshons extend up to infinity. No resistance is offered when Akashon or Prakashon move because all the morphosed ether or anti-ether associated with these packets in very small volume also move. The particles has not to plough through the non existent stationary ether, therefore, there is no resistance. The concept of ether in the present text is not too simplistic in which the ether is considered to be separate from the mass, but in the novel concept all the mass packets and energy packets are considered to be composed of ether shells. The movement of bonded ether is not like the old concept called ether drag where boundary separates the moving object and ether. The bonded ether itself is the moving object; therefore, there is no question of boundary. Akashon (mass packet) and Prakashon (Photon) are compressed states of Akasha and anti-Akasha respectively. The Akasha is perfectly elastic. The displacement of shells leads to the manifestation of cosmic force or the reverse is also true. The unbalanced cosmic force is manifested as field.

If with the novel concept of ether, too many complex theories entangled by net of too complex mathematics can be disentangled as if by magic wand, what is the harm in retaining the concept of ether, though in new form? According to new concept, the existence of ether is very much there. The indirect ubiquitous presence of ether is reflected all around in the manifested creation.

The ether or Aksha is of two types **bonded** and **non-bonded**. The two states are inerconvertible. In case of non-bonded ether, ether shells are merely holes in the ether continuum or Akasha while in case of bonded ether, most of the ether or Akasha of which the shells are composed of is concentrated near the surface of the shell and its density tapers off rapidly as the distance from the shell increases.

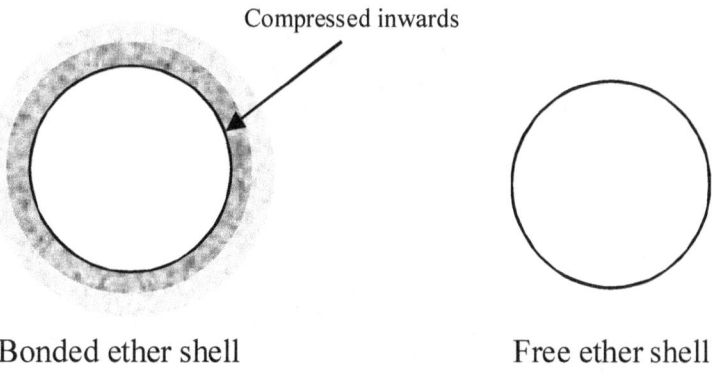

Bonded ether shell   Free ether shell

Fig. 1

The two types of shells are used during particle like and wave like nature, bonded ether dominates during particles like nature while free ether dominates during wave like nature. The radius of the shell as proved later on is equal to 1.1415×10^{-35} m. If the elementary particle is smashed with very strong force, the ultimate building blocks of the matter will be found to be ether shells. Even the quarks are composed of these shells.

USE OF MODELS TO EXPLAIN THE THEORIES OF MODERN PHYSICS:

In the nineteenth century, the natural phenomena were explained by using plain logic and simple models. However, for explaining the twentieth century physics, conventional logic is no more acceptable. More and more bizarre methods and unrealistic concepts are being introduced. The study of physics is becoming more and more like that of higher mathematics. Rejecting the straight path, circuitous and complex paths are being adopted to reach at the same goal. It is not that no realistic physical models can explain complex nature. In this book starting with simple models of mass, etheron (Akashon) and energy, photon (Prakashon), I have explained nearly all the complex theories of modern physics. The representative models of mass and energy in association with plain logical mathematics act as magical key in opening the locked treasure chest holding the secrets of nature. Let us next discuss the process of creation in analytical manner and try to find out the primary mathematical equation to impart quantitative form to the ideas expressed broadly in the general philosophical language of the scriptures.

The process of creation starts with the evolution of conditioned Brahma out of the transcendental Brahma. The conditioned Brahma is governed by the chain of cause and effect while the transcendental Brahma is beyond all comprehensions. The creative agency Maya helps in the birth of conditioned Brahma. The conditioned Brahma is the repository of creation to be manifested as a chicken is present in the egg; therefore, the conditioned Brahma is Brahmand. The conditioned Brahma splits into two parts, Akasha and Anti-Akasha through the creative agency Maya. In fact after the formation of Brahmand, the creative agency Maya dominates the stage. The Akasha does not imply emptiness but it is beyond comprehension in the relativistic sense. The sense of relativistic comprehension is born only in the later stage of creation. The conventional type of subject-object knowledge is not applicable to the understanding of Akasha. The Akasha and Anti-Akasha further split into two parts each of which can be named as the positive and negative parts. The divided state of the Akasha and Anti-Akasha is changed into granular form with the creation of shells.**The shells formed from ether or Akasha are the basic building units. The shells can be named also as cosmic seeds of manifestation.**The four types of shells, two types positive and negative, each of the Akasha and Anti-Akasha merge together to form quadruplets.The formation of the quadruplets and or mayon dissolution to form the undivided Akasha or Anti-Akasha is the continuous and reversible process.

THE BIRTH OF FIELD, MATTER (AKASHON) AND ENERGY (PRAKASHON):

As already stated above, to get musical notes from a wire, the two stage process of stretching and suitable mode of vibration of the elastic wire by plucking is required, similarly for the birth of the manifested creation, the shells of Akasha or Anti-Akasha are displaced or stretched in the first stage (passive motion) and the second stage involves the change of stationary stretched state of the shells into active vibratory motion state. After the completion of the two-stage process, the cosmic symphony of creation is composed or the Brahmand is cracked to reveal the unmanifested state of Brahma as mutable creation along with the primal sound of creation OM. The transcendental Brahma changes into relativistic creation with the attributes of plurality and during the process of creation, active role is played by the creative agency Maya.

The nature of Akasha shells:

The Akasha and anti-Akasha shells are the basic units used for the formation of Akashon (mass packet) and Prakashon (photon) respectively. ($\pm$)Akasha shells are progenitor of mass and gravitation field while ($\pm$) anti-Akasha shells are progenitor of photon energy and electric field. There are four types of shells i.e. positive and negative Akasha ahells and positive and negative anti-Akasha shells. Pure shells of the same type are unstable, hence pure shells combined with other shells to form stable hybrid shells. Pure anti-Akasha shells (charge shells) is unstable, it combines with charge shells of opposite sign to form neutral doublet of charge shells. The stable doublet of charge shells is the building block of photon. The pure charge shell can become stable also by combining with pure mass shell of opposite type to form hybrid charge mass shell. The hybrid charge mass shells are the basic building units of charged elementary particles or conventional matter. The pure Akasha shells are also unstable like the pure charge shells. The pure mass shell can change to stable hybrid shell by combining with pure mass shells of opposite type to a doublet of mass shells. The doublet of mass shells if vibrating can be used to form graviton or if doublet is only stretched dark matter is formed. Dark energy is composed of doublet of charge shells which are only stretched without any oscillations. There are two types of mass and charge. The pure mass and charge are composed of pure mass and charge shells while hybrid or conventional mass charge is composed of hybrid mass charge shells.

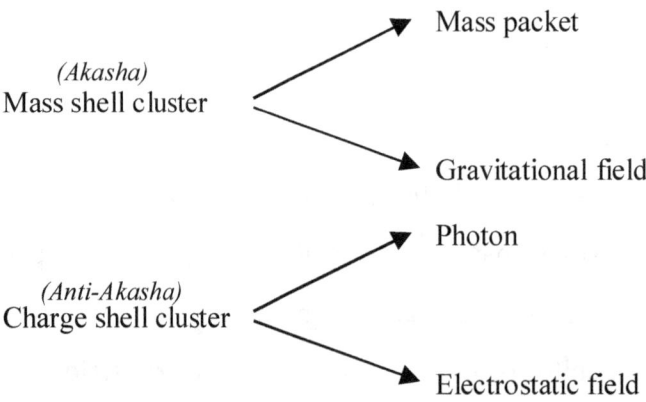

The hybrid state of shells is stable, hence the conventional hybrid matter is composed of hybrid mass-charge shells and is always associated with charge.

The aggregate of four types of shells form unique quadruplet of shells named as Mayon. It is highly unstable and can be called as cosmic seed of creation because in it all the units needed for the creation of phenomenal world are grouped together. The Mayon is not stable because here four units of shells are present as loose combination without hybridization. The properties of Mayon will be discussed later on also.

Types of Properties:

There are two types of properties pure and hybrid. Pure mass packet is composed of hybrid doublet of pure mass shell because doublet is more stable than single shell. Similarly pure charge packet is composed of hybrid doublet of pure charge packets.

The uniform distribution of the shells is disturbed when the shells are stretched away or compressed inwards from the equilibrium positions to form clusters. The displaced shells tend to return to original state when compressed or stretched and thus behave like elastic springs which unlike strings can be compressed or stretched while strings can be stretched only. The displaced shell springsare clustered around a spherical core in which there is no displacement of the shells of the Akasha.

The Cosmic Force:

At the time of creation prior to the conversion of primordial ether (akasha) to morphosed ether as mass and energy packets ether shells were fixed in space The first step in the conversion primordial ether to to morphosed state at time of creation was the displacement of shell centres from equilibrium positions.the Cosmic Force shifted the centres of the shells. The force that displaces the centers of the Akasha shells from the equilibrium position by compressing or stretching is called cosmic force. The cause of the cosmic force is subtle Akasha (ether) or space in the same manner as atmospheric pressure and hydrostatic pressures are the properties of air and water. The cosmic force is manifested with displacement of shells of elastic Akasha.The atmosphere pressure and hydrostatic pressures are not experienced unless the pressure forces are unbalanced. If palm of the hand is placed on the mouth of bottle after sucking out the air, the neck of the bottle sticks to the palm due to the unbalanced pressure acting on the bottle. Similarly, unbalanced cosmic force associated with morphosed ether (Akasha) manifested as field.The cause of all the prime four forces or fields of nature is the unbalanced cosmic force.

The gravitational field is the effect of compressing inwards of shell springs of Akasha. Similarly the electric field is the effect of stretching outwards of the anti-Akasha shellsprings. Form the above discussion, it is concluded that during the stages of creation, the first gross manifestation out of subtle is the evolution of primary or **cosmic force** (field, the Sattva guna) during the displacement of the shells from equilibrium positions to form springs. The matter and energy are manifested as the**cumulative effect of the oscillations (motion-cause) of the displaced shells of the Akasha and anti-Akasha respectively.**

The manifested creation changes to un-manifested state when the primary cause of motion (Maya) ceases to exist and that is why the phenomenal world is said to be illusionary in nature. Shells have two types of transverse oscillations independent and dependent oscillations, the value of which is inversely proportional to the distance of the shell from the geometric centre of the sphere of tranquillity. The independent oscillations are associated with the shells of bonded ether while in case of free ether the center of the shell becomes source of wavelets where each point of the wavelet has same frequency or dependent frequency equal to the frequency of wave originating from shell center. These two types of oscillation will be used to explain dual nature of Akashon and Prakashon.

Evolution of Physical Laws

The sequence of creation is shown in the diagram below

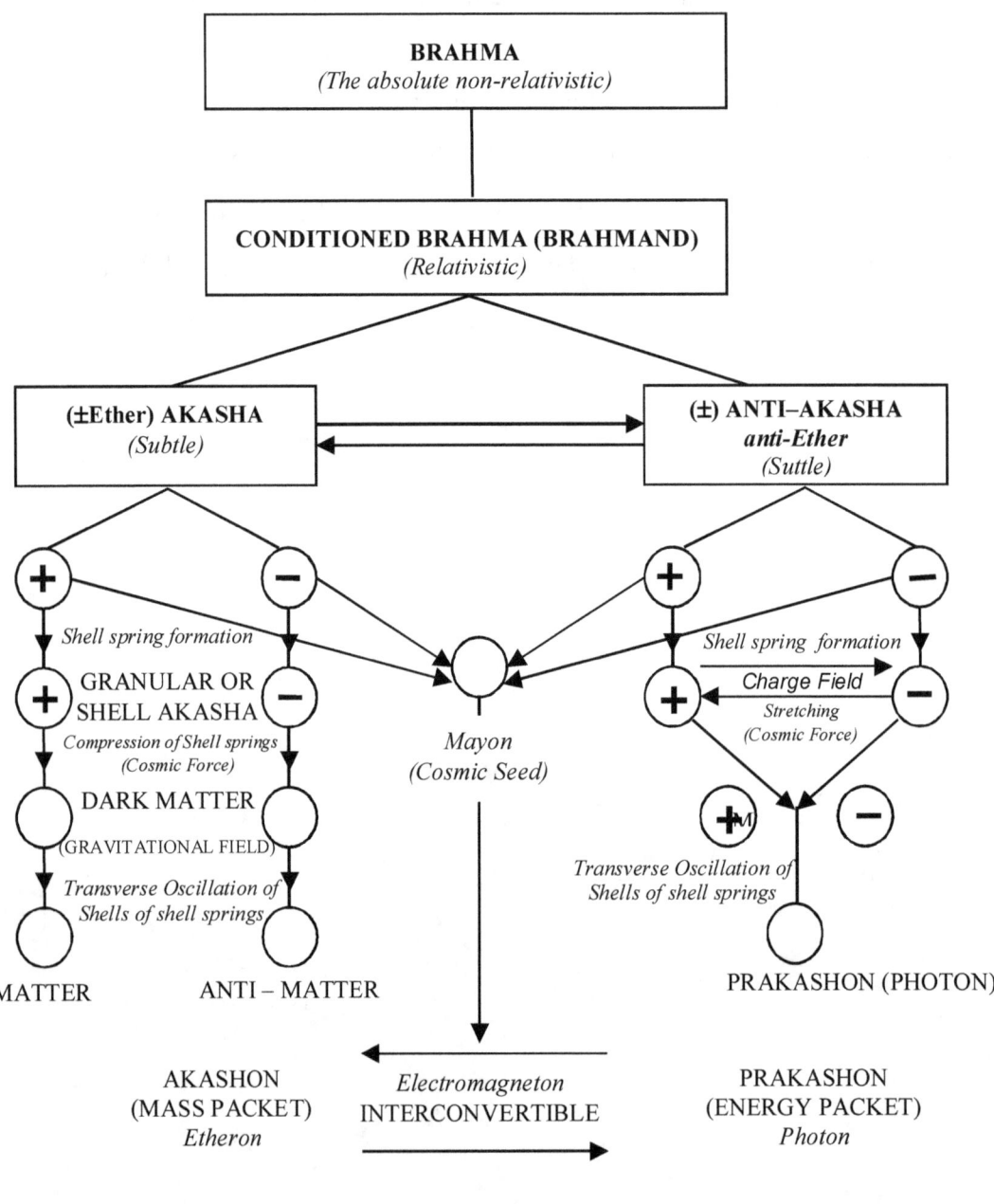

Fig. 2

The **spherical core** of the Akashon (mass packet) or of the Prakashon (photon), energy packet is devoid of any active motion, therefore, it is apt to **name it as the sphere of tranquility of moving Akashon**. The mass centre is located anywhere inside the **sphere of tranquillity** therefore, it is apt to call sphere of tranquility as region of uncertainty. It is seen that representative of mass Akashon or Etheron is not a point as in the classical sense. From above discussion it is seen that ether or Akasha beyond the surface of sphere of tranquillity is associated with the morphosed ether. As the Akashon moves the associated ether or bonded ether also moves with it and hence no resitance is offered to the motion of Akashon. All the properties of the Akashon are associated with the bonded shells. The sphere of tranquillity is filled with primodial akasha.

The space around the sphere of tranquillity in which the displaced shells are oscillating can be suitably called the region of perturbation. The region of perturbation is composed of morphosed ethe(akasha) while sphere of tranquillity is filled with conventional ether .The density of morphosed ether present in the region of perturbation falls very steeply with distance from centre as will be proved later on quantitatively therefore the morphosed ether is practically confined to the surface of sphere of tranquillity The morphosed ether present in region of perturbation with steep fall in density as distance increases is the cause for spooky action at distance and manifestation of field as explained later on The conventional ether present inside the sphere of tranquillity is insulated from outside space. The permanent region of perturbation is composed of bonded ether shells having independent oscillations. In the region of perturbation of

Akashon inwards compressed springs composed of oscillating shells start from the surface of sphere of tranquillity and extend up to infinity are present. In case of Prakashon the springs of infinite length are composed of anti-Akasha shells. One end of the spring is located on the surface of sphere of tranquillity while the other end extends up to infinity, therefore, the second end is loose end which leads to the conclusion that springs do not hamper the free motion of Akashon or Prakashon.

The region of perturbation is limitless, it extends up to infinity as shall be seen later on. The shells are oscillating with the independent oscillations, in other words each shell oscillates with specific frequency unlike the wave propagation where the frequency is same throughout. The displacement of the shell can be either away or towards the centre of the sphere of tranquility. The shells are ultimate building blocks of all the mass and energy. The quarks are also composed of the shells. The shells associated with mass packets (Akashon) and photon (Prakashon) can be exchanged. The radius of the shell as calculated later on is equal to 1.1415×10^{-35}m. No two Akashons (electrons) are identical because the shells of two Akashons executing independent oscillations are not in the same phase. If special techniques are used for the production of photon and electrons in which all the shells are in the same phase, these special photons or electrons are said to be entangled state. The entanglement is there even if the distance of separation is large. It is so because the region of perturbation is limitless.

Akasha shells line up to form springs, which are stretched outwards or inwards.

Primordial akasha:

It was that akasha which was present before creation; after creation the primordial akasha was confined to sphere of tranquillity and hence it is not present in free space and that is why Michelson–Morley could not detect it. It is used for the formation of temporary region of perturbation inside sphere of tranquillity and exhibition of wavelike property when it is ejected out.

Morphosed akasha:

It is that akasha which is manifested as creation. Primordial akasha metamorphosed to morphosed akasha when at time of creation primordial akasha shells were subjected to cosmic force and oscillation prcess to be manifested as gross form of the subtle form The region of perturbation is composed of morphosed akasha with rapid descrease in density with distance from centre of akashon or prakashon hence space is filled with very attenuated presence of morphosed akasha Spooky action at distance takes place through morphosed akasha as explained later on.

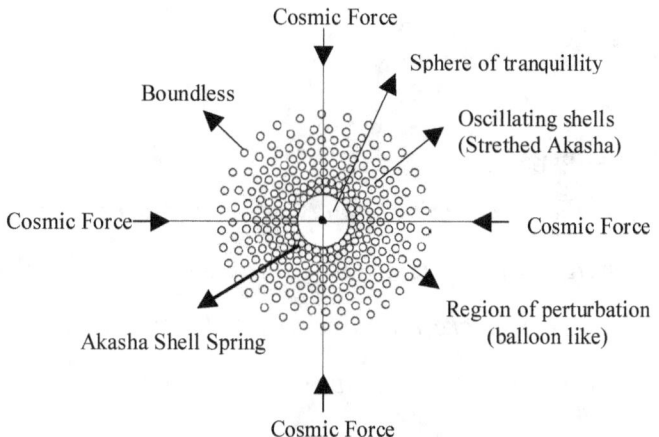

Fig. 3(a). Akashon (Etheron - mass packet, Electron or Proton)

Evolution of Physical Laws

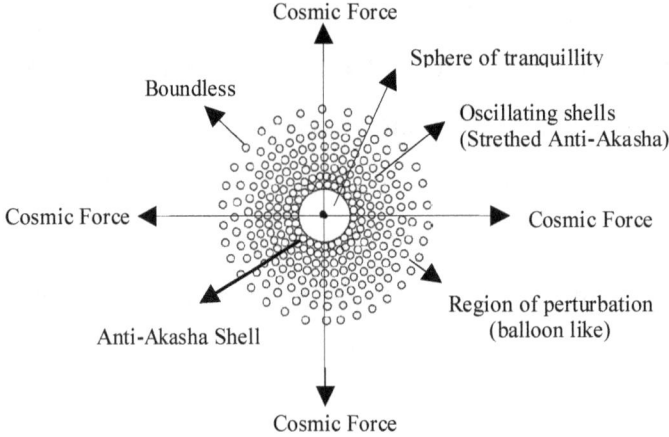

Fig. 3(b). Prakashon (Photon)

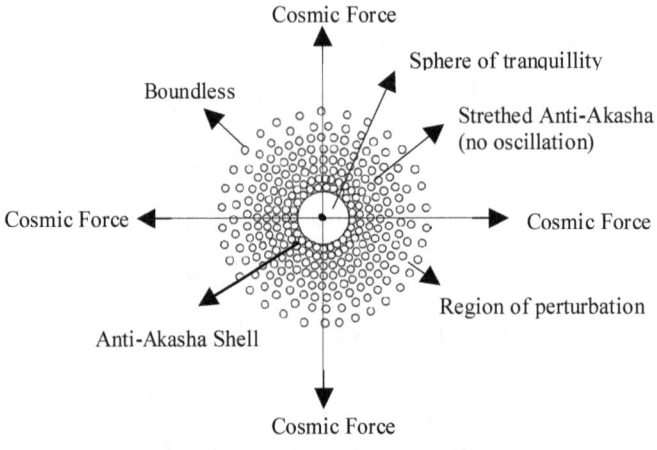

Fig. 3(c). Charge Packet

What are mass and energy:

Mass and energy are the cumulative effects of oscillations of compressed shells of shell springs of Akasha or stretched anti-Akasha shells respectively. Mass and energy are the condensed state of Akasha and anti-Akasha. The convenetional matter is composed of stretched and oscillating hybrid mass-charge shells.

What is charge packet:

In charge packet outward stretched springs of anti-Akasha shells are present without any shell oscillations. When shells oscillate charge packet changes to energy packet.

The proposed structure of mass packet and photon can explain the point-like and extended size like behaviour of these particles.

DERIVATION OF MOTHER OF ALL EQUATIONS:

The boundary separating the region of perturbation and the sphere of tranquillity can move inwards or outwards, in other words the radius of the sphere of tranquillity can increase or decrease but not continuously because shells with finite size are to be accommodated. When there is a change in the volume of the sphere of tranquility, there is a corresponding change in the gross creation (region of perturbation) because with change of volume of sphere of tranquillity the primordial ethrer present in the sphereof tranquillity changes to morphosed state and is thus transferred to region of perturbation. The above statement can be expressed in mathematical terminology in very simple way. The volume of the sphere of tranquillity depends on the radius and therefore the magnitude of the property (mass, energy) also depends on the radius of the sphere of tranquillity.

The rate of change of the volume of sphere of tranquillity and the magnitude of the gross property with change in the radius of the sphere of tranquillity are inversely proportional. The radius of sphere of tranquillity is equal to the distance of the center of the innermost shell of the shell springs from the center of sphere of tranquillity.

If V, x and r are the volume, magnitude of the property and radius of the sphere of tranquility, the rate of change of volume is dV/dr and that of magnitude of the property is dX/dr

$$dV/dr \propto 1/(-dX/dr)$$

The negative sign is put before dX / dr because when V increases x decreases and vice-versa.

dV / dr (-dX / dr) = k_x = universal constant

The above equation is the fundamental equation which is useful in deriving quantitatively many useful laws governing the phenomenal world. The abstract philosophical thoughts related to creation are expressed in mathematical manner in the above simple equation. The beauty of the truth lies in its simplicity. It is not always necessary that to unravel the mysteries of nature complex mathematics is required. Our goal is to reach the truth, either by following the direct path or choosing the complicated way.

The gross property x is the magnitude of the property due to cumulative effect of the oscillating shells or it may also be property associated with the single shell.

The differential form of the fundamental equation can be changed into integrated form

$V = 4\pi r^3 / 3$

=> $dV / dr = 4\pi r^2$

When, r → ∞, there is no region of perturbation, therefore; x → 0

=> $\int_{X}^{0} dX = -(k_x / 4\pi) \int_{r}^{\infty} dr/r^2$

=> $4\pi r . X = k_x$ = universal constant

In the above integration, the lower limit of r is not zero because r cannot be less than radius of the shell; therefore, there is no singularity during integration. Singularity is introduced if size of the Akashon is taken equal to zero as is the usual convention, but actually as we have seen, the size of the Akashon extends up to infinity with the condition that during integration lower limit r is not zero. There is no need to introduce artificial concept of renormalization to avoid singularity if the above model is adopted. The above equation can be called as the principle of inverse variation because x and r are inversely proportional to each other. This equation can also be named as *'Mother of all equations'*. The mass and energy associated with the Akashon the mass packet (mass packet) and Prakashon (photon) can be expressed in mathematical equation. Akashons or Etheron, the representative of mass are synonymous.

$4\pi r_m . m = k_m \rightarrow$ mass equation

$4\pi r_E = k_E \rightarrow$ energy equation

The mass equation points towards the existence of black holes because when $r \rightarrow 0$, $m \rightarrow \infty$, but actually r is never zero, but it can be very very small, hence m can be very huge.

With change in the nature of the property X, there is change in the universal constant associated with the equation. The constants associated with the equations are independent of the motion.

The equation associated with the principle of inverse variation is general in nature and is applicable to know the magnitude of other properties also. In addition to the mass and energy which are cumulative effects of oscillations of shells, the principle of inverse variation is applicable to know the magnitudes of other properties such as angular momentum, magnetic moment, electromagneton energy, etc. If L_s is the angular momentum of Akashon and r is the radius of sphere of trqnauillity $4\pi r L_s$ = constant. The constant should have dimensions of (angular momentum) x (length) = h.r. Here h is Planck's Constant and r can be taken equal to the radius of sphere of tranquillity of the given Akashon hence is also constant.

$$L_s = h/4\pi = \tfrac{1}{2}(h/2\pi)$$

The simple calculations give the angular momentum of Akashon and spin.

The principle of inverse variation is also applicable to know the magnitude of properties associated with the individual shells such as frequency, displacement etc. If v is the frequency of oscillation of the shell located at distance r from the center of the sphere of tranquillity

$4\pi r.v = k_v$ = constant
Constant k_v has got dimentions of velocity.
If t is the time period of oscillation, $v = 1/t$
$4\pi r/t$ = invariant constant

r/t has got dimensions of velocity, therefore r/t or 4πr/t is unique velocity which is invariant. Each oscillating shell can be considered to be source of unique waves having unique invariant velocity. The concept of invariant velocity is logical natural consequence of principle of inverse variation.

Similarly, if d is the displacement of the shell
$4\pi r.d = k_d = $ constant
If N is the limiting number of displaced shells present in the region of perturbation
$4\pi r.N = k_N = $ constant

Minkowski Space Equation can be got from the principle of inverse variation with little manipulation as will be seen later on.

The other properties also which are associated with the Akashon such as self electrical potential energy, permittivity, angular momentum, magnetic dipole moment which are calculated by applying the principle of inverse variation are called natural properties. The first *guna* of *Maya* to be created was the *Sattva guna*, the unifying and the balancing principle. The *Sattva guna* stands for the force or field.

Each stage of the creation process viz. atomization of the Akasha to from shells, displacement of the shells and imparting oscillatory motion to the displaced shells required special force. The special force is named as the **Cosmic Force** or the **Primary Force**. When the Akasha changes into granular from, the Tamas guna (inertia) of Maya is born, when the Sattva and Tamas gunas participate jointly in the next stage of creation to impart oscillatory motion to the shells the effect of which is the gross creation, the Rajas guna comes into existence. The time is space dependent similarly, energy and mass depend upon radius of sphere of tranquillity therefore, like time mass and energy can be considered as fourth dimension.

Extended Size of Akashon:

The Akashon has definitely got extended dimensions contrary to the paradigm of point like existence. The Akashon is not merely enigmatic blob. All that is manifested in nature whether at macro or micro level has got internal structure based on regular pattern. Once the internal structure is known, it becomes very easy to understand and demystify the properties of system under study. If the dimensions are point like, no internal structure can be proposed. The elucidation of structure of molecules revolutionized the srudy of chemistry and the present day chemists are not groping in the dark while trying different combinations to get a product of desired properties.

The sub-atomic particles cannot be assumed to be point like. How the stable superstructure can be raised if the foundation is shaky. Once the internal structure is known it becomes very easy to predict the laws of nature based on the interaction of the sub-atomic particles. The modern theories of physics can be understood by the application of evergreen classical mechanics. Starting with the extended size concept and assigning regular structure to the particles, the study of complicated theories like relativity field quantum mechanics etc. becomes child's play as will be seen later in the pages of this text. It becomes also very easy to understand the concept of mass and energy. In the proposed structure of Akashon and Prakashon, the mass and energies are distributed over the region of perturbation (extended size) but at the same time mass and energy can also be supposed to be concentrated at the center of Akashon and Prakashon (point size).

Entanglement:

While deriving the principle of inverse variation it was seen that the region of perturbation of Akashon and Prakashon are limitless. The extended dimension concept easily leads to the conclusion that not only the photon but all that is manifested exist in state of entanglement. Nothing is isolated. The spooky action at a distance is possible because region of perturbation is limitless. During entanglement locality rule is never violated because even when the particles centers are separated by huge distance they are still in touch. If the dimensions are taken as points, no entanglement is possible.

Types of properties of Akashon:

There are two types of properties of Akashon – pure properties and blended properties. To discuss the above point we can take examples of mass and charge. Uptill now we have been hearing of only one type of mass, but actually there is blended mass and pure mass. As the name applies in the case of pure Akashon, the oscillating and stretched shells of Akasha are present while in the blended mass hybrid shells of Akasha (mass effect) and anti-Akasha stretched shells (charge effect) are present. The blended mass packet is also an example of blended charge packet. The above explanation does not imply that neutral mass is pure mass. The mass packet can be neutral also due to the presence of equal and opposite type of stretched charge shells. But here the pure mass packet applies that only mass shells are stretched. Similarly, in the case of pure charge packet only stretched charge shells are present while in the blended charge packet, stretched mass shells are also mixed with the mass shells or the hybrid stretched shells are present. The mass packet (electron, proton) are hybrid in nature because of hybrid charge while photons exist in pure charge state as will be discussed later on.

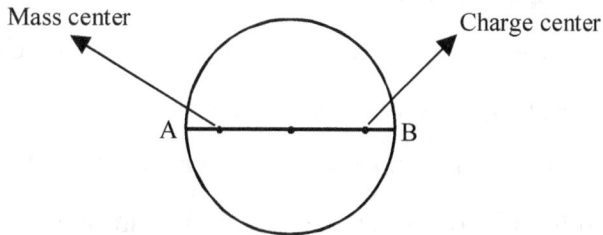

Fig.4. Hybrid Mass-Charge Shell

Evolution of Physical Laws

THE FORCE AND ENERGY:

The extended size of Akashon (mass packet) is needed to know force and energy. Each shell present in the region of perturbation is squeezed inwards or stretched away from the center of the sphere of tranquillity. The inwards compressing or outwards stretching of the shells of the springs depends on the nature of the property manifested by the cluster of shells. The inwards compressing is akin to the atmospheric & hydrostatic pressure experienced by the terrestrial and aquatic animals. The *Sattva guna* of the Maya, the Cosmic Force is required for shifting the position of the shells in the above-mentioned manner or it can also be said that it is the displacement (passive motion) of the shells which is the birth of the *Sattva guna* (Cosmic Force). If uniform distribution of Akasha around shell is disturbed by subjecting it to outside force or by interaction mass center of shell shifts in a direction opposite to the direction of force to counter the impressed force, which proves that action and reaction are equal and opposite (Newton).

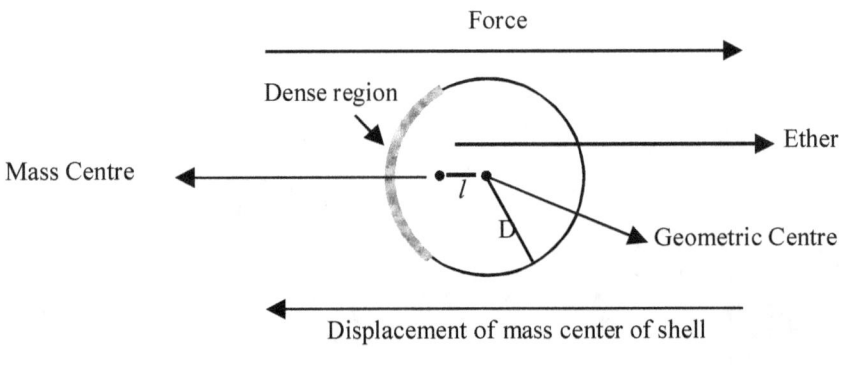

Fig. 5

Mass center is the center of distribution of Akasha around shells. Suppose l is the distance of separation of mass center and geometric center of the shell of radius r, the force acting on the isolated shell due to disproportionate distribution of Akashon is given by Hooke's Force Law

Force = $F_0.(l/r)$

When Akasha shells combine to form Akashon same force law is applicable to the force acting on Akashon.

In case of an Akashon (mass packet) the Cosmic Force acting on each shell is towards the center of the sphere of tranquility. If the isolated Akashon is supposed to be divided into two halves it is easy to see that the resultants of the Cosmic Force acting on the shells present in each half are equal and opposite, therefore, no net Cosmic Force acts on an absolutely isolated Akashon (mass packet).

There are two centers of the Akashon (mass packet). First center is the geometric center of the sphere of tranquillity and the second center is the point at which the resultant of the Cosmic Force acts if there is any. In case of an isolated Akashon (mass packet) at rest or uniform motion, the two centers overlap. The geometric centre is associated with the centre of the shells and mass centre is associated with the distribution of Akasha on the surface of shells. The second basic guna of the creative agency Maya is the Tamas guna which stands for inactivity. The first basic Sattva guna tends to bring change while the Tamas guna opposes it; Thus from the non-dual Brahma, Maya the creative agency having dual nature is born.

Evolution of Physical Laws

Let us interpret the Tamas guna in the scientific terminology of the phenomenal world. When the uniform distribution of Akasha around the shell surface is disturbed each shell experience force directed away from the dense region.

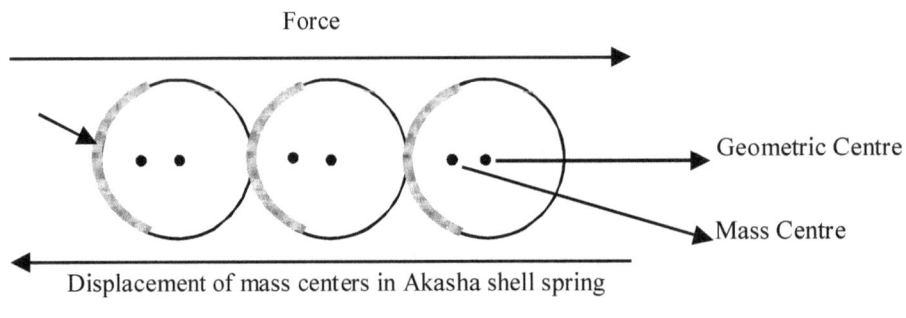

Fig. 6

In case of Akashon (mass packet) the shells formed out of the Akashon are squeezed inwards towards the center of the sphere of tranquillity with Cosmic Force or it can be said that the Cosmic Force is born with the displacement (passive motion) of the shells of Akasha. The principle of inverse variation is applicable to determine the magnitude of the displacement of the mass shell center. If dr is the displacement of the mass shells center located at distance r from the center of the sphere of tranquility, the displacement is given by the following equation –

$$4\pi r.dr = k_d \quad => \quad dr = k_d / 4\pi r$$

The Cosmic Force acting on each shell is proportional to the displacement. The Cosmic Force = $F^* = -k^*.dr$. the relation is almost similar to the familiar Hooke's law i.e. $F = -k.dr$, which is applicable to the bulk gross form of the creation but with the difference that while in the conventional relation the constant of proportionality k is almost constant, in case of Cosmic Force relation, the value of k* depends on the distance of the shell from the center of the sphere of tranquility. It is evident that for the shells situated near the surface of the sphere of tranquility, the value of shell density is high, therefore, the high value of k* is expected to be high. In other words the principle of inverse variation is applicable to determine the magnitude of k*

$4\pi r . k^* = k = $ constant

Putting the values of k* and dr in the Cosmic Force relation

$F^* = -(k.k_d / 4\pi).(1/ 4\pi r^2) = -k^*_F / 16\pi^2 r^2$, where $k.k_d = k_F^*$

Multiplying both sides by dr

$\Rightarrow dw^* = F^*.dr = (-k^*_F / 16\pi^2 r^2).dr$

w* is new property introduced here

The property F*.dr in magnitude is associated with each shell which is simply displaced in the region around the sphere of tranquillity prior to acquiring the oscillatory state.

There are three active steps involved in the process of creation –
1. The atomization of Akasha and anti-Akasha with the formation of the shells of ether foam. The process is reversible.
2. The displaced shells form clusters or springs around the sphere of tranquillity and during the displacement stage of the shells, the Cosmic Force or field is manifested.

Evolution of Physical Laws

3. Subsequently the independent oscillation of the displaced shells is the active cause (Maya) of creation and the effect is manifested as the gross matter or energy. It is in the third stage of creation the derived guna Rajas of Maya is manifested.

The third step of creation is very helpful in calculating the magnitude of the mass and energy associated with the Akashon (mass packet) and Prakashon (photon) respectively.

As already discussed, during the second step of creation the unnamed property $dw^* = F^*.dr$ is pumped into each shell during the displacement process. If a line is imagined to be originating from the center of the sphere of tranquillity of the Akashon or Prakashon which extends upto infinity, the numerous displaced shells can be spotted on the line with gradual increasing order of displacement when the direction of locating the shells is towards the center of sphere of tranquility. The increase in displacement can be considered as continuous, therefore, the total magnitude of property w^* associated with the shell when it is brought from infinity up to the point under study can be determined by integration

$$w^* = \int dw^* = -k_F^* / 16\pi^2 \int_{\infty}^{r} dr/r^2$$

$$\Rightarrow w^* = k^*_F / 16\pi^2 r$$

$$\Rightarrow 4\pi r.w^* = k_F^* / 4\pi = \text{constant}$$

It is seen that the principle of inverse variation is applicable to determine the magnitude w* of un-named property associated during the displacement process in the second stage of creation. The property w* is natural or fundamental property.

The property w* associated with each shell during the second stage is still in the unmanifested stage of creation. The Akashon (mass packet) and the Prakashon (photon) are manifested only when the displacement acquires independent oscillations. With mere stretching of the Akashon shells and anti-Akashon shells, gravitational field and charge fields are manifested respectively. The cumulative effect of independent oscillations of the stretched shells of the Akasha or anti-Akasha present in volume $4\pi r^2.dr$ at distance r from the center is the manifestation of mass dm or energy dE of Akashon or Prakashon. The principle of inverse variation is applicable to determine the magnitude of property w*, therefore, it is one of the fundamental properties. It is named as energy. **It is seen that magnitude of enrgy can be determined by the product of force and the distance through which the force applied point moves.** The energy calculation formula is not fixed arbitrarily but here the reason for expressing it as such is also clear because **energy is natural or fundamental property obeying the principle of inverse variation.**

When associated with the displaced shells without any oscillatory motion, it is called mutable energy because in that special state, it can be manifested as Akashon (mass packet) or conventional energy (Prakashon, energy packet). The units of mutable and conventional energy are the same, that is, the product of force and displacement. From the above discussion it is easy to conclude that the Akashon (mass packet) can change into Prakashon (energy packet) through the intermediate mutable energy packet.

Akashon	↔	Mutable energy packet	↔	Prakashon
(mass packet)		(Mattenergon)		(energy packet)
(etheron)				*(photon)*

Mattenergon is coined by combining matter and energy. Mattenergon is a hybrid of matter and energy with dual nature.

The mass is cumulative effect of all the shells present in volume $4\pi r^2.dr$. Suppose dm is the mass present in this volume, the mass dm is proportional to energy dw*.

$dm = k_0.dw*$

putting value of dw*,

$dm = (k_0.k*/16\pi^2 r^2).dr$

on integration, $m = \int_0^m dm = -(k_0.k*/16\pi^2).\int_\infty^r dr/r^2$

$4.\pi.r.m = k_0.k*/4\pi = k_m$

The above is the same mass equation as got by applying the principle of inverse variation to get the mass of Akashon or mass packet. Similarly the manifestation of Prakashon (photon) can be mathematically discussed. If w* is the mutable energy associated with the single shell at distance r and w_E is the magnitude of manifested conventional energy, by following the same steps as in the case of the determination of the magnitude of the mass of an Akashon, the magnitude of energy of the Prakashon can be calculated as dE is the energy present in volume $4\pi r^2.dr$, its magnitude is proportional to energy associated with single shell.

$dE = k_0'.dw*'$

putting the value of $w*'$ and limits

$$E = \int_0^E dE = -(k_0'.k*/16\pi^2).\int_\infty^r dr/r^2$$

$4.\pi.r.E = k_0'.k*'/4\pi = k_E$

The above is the energy equation of photon or energy packet.

The relations derived for calculating the mass energy associated with the Akashon and the Prakashon are the same as got by the application of the principle of inverse variation.

The mass and energy relations for the Akashon (mass packet) and the Prakashon (energy packet) can be derived in another simple manner also. The mass and energy are the cumulative effects of the oscillations or perturbation of the displaced shells present in the region of perturbation of the Akashon and Prakashon, therefore the change in the magnitude of mass and energy depends on the change in the number of dNshells which are present in the volume $4\pi r^2.dr$ created as a result of radius changed by dr. The total number of shells is given by

Evolution of Physical Laws

$4\pi r N = k_N$

$\Rightarrow dN = - k_N . dr / 4\pi r^2$

if dm is the change in the mass as a result of change in the number,

dm α dN

$\Rightarrow dm = k_0 . dN$

$\Rightarrow dm = - (k_0 . k_N / 4\pi).(dr / r^2)$

$\Rightarrow \int_0^m dm = - k_0 . k_N / 4\pi \int_\infty^r dr/r^2$

$\Rightarrow m = k_0 . k_N / 4\pi r = k_m / 4\pi r$

$\Rightarrow 4\pi r m = k_m$

Similarly, the relation can be derived for the magnitude of the energy associated with the Prakashon

$dE \propto dN$

$\Rightarrow dE = k_0' . dN$

$\Rightarrow dE = - (k_0' . k_N / 4\pi).(dr / r^2)$

$\Rightarrow \int_0^E dE = - k_0' . k_N / 4\pi \int_\infty^r dr/r^2$

$\Rightarrow E = k_0' . k_{N'} / 4\pi r = k_E / 4\pi r$

$\Rightarrow 4\pi r E = k_E$

There is lower limit to radius of the sphere of tranquillity because when the surfaces of two shells located diametrically opposite on the surface of sphere of tranquillity touch each other, further reduction in the radius of the sphere of tranquillity is not allowed, therefore, the lower limit of the radius is equal to the radius of shell which is equal to 1.1415×10^{-35}m. At the time of big-bang it might be that radius of shells was much smaller than the above value, hence accumulation of very large amount of mass was possible in nearly point volume. The above observation shows that magnitude of the property calculated by applying the principle of inverse variation is finite, it cannot approach infinity or there is no singularity because radius cannot be equal to zero. The mass and energy are the effects of primary cause of perturbation or the Maya. The above relation for the magnitude of mass and energy are derived by relating cause and effect with the help of mathematical equations.

NO OBSERVATION OF SINGULARITY:

When the magnitude of properties associated with the region of perturbation (mass energy etc.) is calculated the lower limit of integration is not zero but, it is equal to the radius of sphere of tranquillity, therefore, the magnitude of the properties is never infinite but it is finite. Therefore, there is no need to bring in the complicated mathematical tool of renormalization to bring down the magnitude of properties within limits.

THE FORCE AND INERTIA:

As already seen the birth of Cosmic or Primary Force and the displacement of the Akasha are intimately related. The shells are displaced towards the center of the sphere of tranquillity in case of an Akashon. The magnitude of cosmic force acting on each shell is proportional to the displacement of the shell center. The above displacement related way of measuring the force can be called static method.

Let us next discuss how the force is related to the motion of the Akashon as a whole. When the mass center (force center) and the geometric center of the Akashon overlap, no net force acts on the Akashon. Suppose somehow either by interaction or direct contact, the shells mass centers in the left portion of the region of perturbation are displaced more towards left because bonded Akasha distribution on left part of Shells is more denseas compared to the shells present in the right portion. The unequal distribution of theAkasha in the two portions of the region of perturbation makes left region more dense as compared to the right portion and hence the mass center shifts away towards left from the geometric center. The overall geometric shell center displacement is zero; therefore there is no shift in the geometric centre.

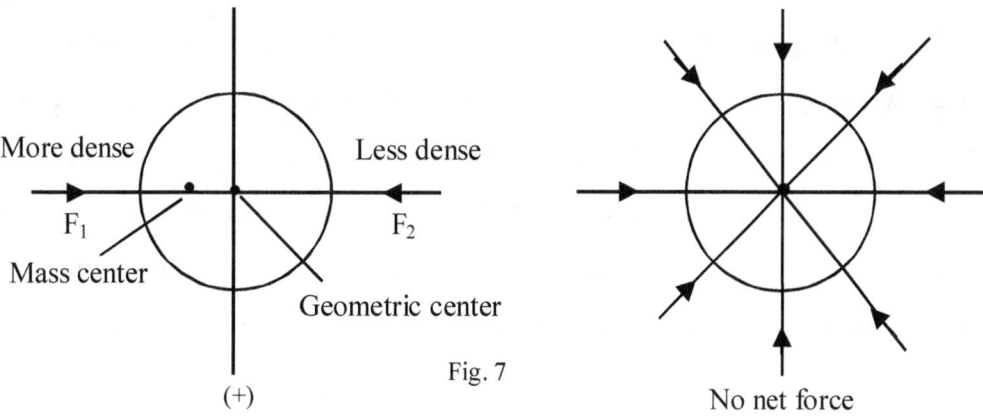

Fig. 7

Evolution of Physical Laws

As a result of unequal displacements of shellsmass centers, the net resultant forces experienced by the two halves of the region of perturbation are not equal. The force F_1 from left to right experienced by left part is more than the force F_2 from right to left experienced by the right portion. The force F_1 is more than F_2, therefore the Akashon experiences net force in the direction from left to right. The net force can be named as the secondary force. The inwards squeezing of the shells or springs is not an instantaneous process. The shells or ether resist further change in the position; therefore, the shifting away of the mass center away from the geometric center is also not abrupt. As a result of redistribution of the ether, the mass centre of each shell and mass centre of Akashon as a whole is shifted away from the geometric centre. During shifting away of the mass exerts force opposite to the external force and the Akashon does not move. This shows action and reaction are equal and opposite. The time interval during which mass center is shifted away to the maximum extent is called inertial time interval. After the maximum displacement has been achieved, the opposing force becomes less than the impressed force and Akashon or etheron starts moving. The Akashon resist change in its state of rest or motion. **The time interval during which the Akashon does not experience the change of state of motion in spite of the force acting on it is called inertial time interval and this property of Akashon to resist the change is called inertia *(Tamas guna)*.** The mass center moves away from the geometric center during the inertial time interval but the Akashon or geometric center as a whole does not experience any change in its state of motion. Change of state of rest into motion or change in motion is not instantaneous process.When the mass centre is merely being separated without the

Akashon undergoing any motion as a whole the work done leads to formation of temporary inertial mattenergon, which remains associated with the Akashon so long as it is moving. The potential energy of mattenergon changes to kinetic energy. The term mattenergon is coined by combining matter and energy hence it is hybrid in nature having dual character. The shell by shell formation of inertial mattenergon also needs some time. As the mass center moves away from the geometric center energy is liberated which creates mattenergon or temporary region of perturbation. The extra created surface of sphere of tranquillity moves inward while mass center moves outward toward the surface of sphere of tranquillity during the inertial time interval. It will be seen later on that mattenergon plays important part in the study of quantum mechanics and uncertainty principle. Mattenergon has got sphere of tranquillity with same center as that of the Akashon with which it is associated. **The inertial time can also be said that time period during which mattenergon is created and the shells are rearranged.** As the mass packet starts moving with the uniform velocity, the mattenergon center starts oscillating with geometric center as mean position. Mattenergon has got dual nature. Mattenergon is hybrid state of matter and energy. Mattenergon is of two types, inertial mattenergon and motion mattenergon. Inertial mattenergon is created during inertial time interval and motion mattenergon is created when Akashon is moving as a whole. Motion mattenergon is composed of inertial mattenergon. Mattenergon is real in nature not a virtual particle. Mattenergon has no independent existence. It carries energy of temporary region of perturbation and co-exist with moving invariant mass Akashon.

The already displaced shells are squeezed inwards more easily during the initial stages but as the squeezing in process continues, the rate of displacement of shells slows down and consequently the rate at which the mass center moves away from the geometric center becomes less with time and after the inertial time interval the rate of separation is zero. **It can be said that the mode of moving of the mass center away from the geometric center is equivalent to simple harmonic oscillator because F $\propto$ displacement.**

The Akashon or etheron experiences change in state of motion after the expiry of the inertial time interval and at this stage the Akashon experiences the same acceleration as the separated mass center shifted away to the maximum is subjected to towards the geometric center. As the motion of the mass center is similar to the point executing simple harmonic motion, the acceleration at the extreme position is given by, $a = -\omega^2 l$. If the Akashon was initially at rest, it will start moving with acceleration 'a' and require velocity $v = a.t_0$, where t_0 is the inertial time interval and during this time the mass center will reach the mean postion. If no further force is applied the Akashon willcontinue to move with uniform velocity v. The Akashon will be having more mass because during time t_0 work is done, extra region of perturbation is created inwards to accommodate energy or mattenergon. If further force is applied now to the moving Akashon the above steps are repeated. It can be said that extra or temporary region of perturbation is created step by step and it is composed of many inertial mattenergons.

Here l is equal to the maximum distance of separation and ω is angular frequency. During the inertial time interval t_0, one fourth of oscillation is completed, therefore, total time interval $T = 4t_0$

$\omega = 2\pi / T = \pi / 2t_0$
$l = - a / \omega^2 = - (4 / \pi^2).at_0^2$
neglecting the negative sign,
$l = (4 / \pi^2).at_0^2$

The principle of inverse variation is applicable to determine the value of inertial time interval.

$4\pi r_0 t_0 = k_t = $ constant

r_0 is the radius of the sphere of tranquillity of the Akashon (mass packet) at rest before it just starts moving. The radius of sphere of tranquillity becomes less than radius r_0 with the application of force because mass center moves away from geometric center, energy liberated is used to built up extra temporary region of perturbation of moving Akashon. If the force impressed is not large the change in radius can be neglected. If the time interval of the application of the force on the Akashon is less than the inertial time interval, the Akashon does not move as a whole, the force simply brings about the separation of the mass center and the geometric center. When the Akashon (mass packet) in state of motion experiences secondary force, the same result as applicable to the stationary Akashon are got but with the difference the inertial time t for the moving Akashon is

Evolution of Physical Laws

given by the relation $4\pi rt = k_t$. Here r is the radius of the sphere of tranquillity of the moving Akashon just before it experiences acceleration after inertial time interval t.

Inertial time interval and unique Akashon:

There is limit to the maximum mass associated with the Akashon. The Aakashon having maximum mass is called unique Akashon or N Akashon. The radius of the unique Akashon is minimum and is equal to the radius of the shells of which it is composed of. The radius has got minimum value because further decrease is not possible because surfaces of shells located on the opposite ends of diameter just touch in limiting case. In unique Akashon the radii of shells and radius of sphere of tranquillity of Akashon are equal.

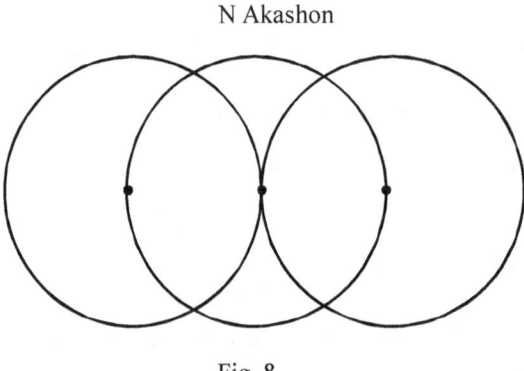

Fig. 8

Suppose the force is impressed on the Akashon. The mass center will start moving away from the geometric center till it just touches surface of sphere of tranquillity. The Akashon will not move because further addition to the region of perturbation is not allowed. When the mass center just touches the surface, the oscillating shell located on the surface must be present in zero force state or mean position. In other words during the time mass center just touches the surface the shell must complete half of oscillation. The frequency of shell on the surface is given by $4\pi r_0 v_0 = c$. The time required to complete half of oscillation = $2\pi r_0/c$. The above time interval is alsoi equal to inertial time interval for the unique Akashon. The inertial time interval for Akashons is given by the equation

$4\pi r_0 t = k_t,$ $t = 2\pi r_0/c$
$k_t = 8\pi^2 r_0^2/c$

Inertial time interval for unique Akashon t_0 equal to $2\pi r_0/c$ is the minimum inertial time interval because r_0 is the minimum.

When an Akashon having mass less than the unique Akashon is being subjected to increasing force the extra mass acquired due to shifting away of the mass center will go on increasing. The stage will be reached when the mass center just touches the extra created surface of sphere of tranquillity. The above state is state of maximum force and the mass of the Akashon is equal to N Akashon. The above observation shows that the same **maximum force** can be impressed on the Akashon irrespective of the initial mass.

CREATION OF TEMPORARY REGION OF PERTURBATION:

When the oscillating shells are displaced more inwards by the secondary force, the conventional energy is pumped into the shells and to accommodate extra energy, more shells are created on the inner side of the surface of the sphere of tranquillity. In other words extra region of perturbation is created when the Akashon is set into motion, i.e. after the inertial time interval. The extra region of perturbation creaed during motion is composed of mattenergy or hybrid shells and hence, it is temporary in nature. Hybrid shells are composed of bonded and free ether shells. Permanent region of perturbation of Akashon is composed ofusually hybrid mass charge ether shells.

The extra region of perturbation grows inwards contrary to the conventional growth which is outwards and the extra region of perturbation is temporary in nature. The radius of the sphere of tranquillity of the moving Akashon is less than that of the stationary Akashon due to the extra inwards growth of the temporary region of perturbation. The mass associated with the permanent region of perturbation is the invariant mass. The temporary region of perturbation is composed of hybrid shells of free andbonded Akasha; therefore, it has got dual character of particle and wave. What is stored in the temporary region of perturbation can be called as mattenergy. The hybrid shells are composed of hybrid ether. The ether or anti-ether is of two types, bonded and free. In case of bonded state the ether or anti-ether is concentrated in the surface of the shells and hencewhen the shell vibrates no wavelets start from the shell but in case of free ether the shells are merely holes in the ether continuum. When the free ether shell vibrates the shell is source of oscillations because free ether is present around the shell to carry away the waves. The hybrid ether exhibits properties of free ether or bonded ether one at a time. The temporary region of perturbation is composed of hybrid ether. **The temporary region of perturbation is formed when the mass packet is in motion therefore, mass packet in motion or energy packet will exhibit dual nature particle like or wave like. It is the formation of temporary region of perturbation which imparts dual character to mass packet in motion** which shall be studied later on. The growth of temporary region of perturbation is not indefinite because further scrutiny of behavior of mass packet discussed later on will reveal that there is limit to the maximum force which can be applied to the given Akashon or mass packet. The mass associatedwith permanent region of perturbation is invariant mass and mass associated with temporary region of perturbation can be called rider's mass or mattenergon because it is riding on the invariant mass. The sum of these two masses is relativistic mass. The sphere of tranquillity is filled withconventional ether which is used to form temporary region of perturbation.

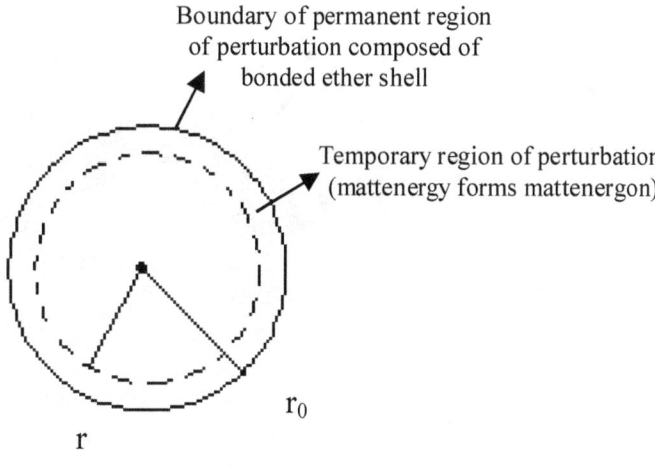

Fig.9 Moving Akashon

Temporary region of perturbation is composed of hybrid Akasha i.e. it contains bonded and free Akasha and hence here in lies the dual nature of Akashon in motion. **No motion, no temporary region of perturbation and hence no dual character.** The temporary region of perturbation is also repository of extra energy generated as a result of motion.

DETERMINATION OF THE CONSTANTS ASSOCIATED WITH THE EQUATION OF THE PRINCIPLE OF INVERSE VARIATION:

The following three equations associated with the mass, energy and frequency of the shells are written by using the principle of inverse variation or mother of all equations.

$4\pi rm = k_m$ = mass constant

$4\pi rE = k_E$ = energy constant

$4\pi r\nu = k_\nu$ = frequency constant

The constants associated with the principle of inverse variation are universal constants.

$E/m = k_E / k_m = k$

The dimensions of k will depend on how the mass and energy are expressed. The energy relation has been derived not assumed as,

Energy = Force x Distance

The force relation has been derived in the next topic

Force = Mass x Acceleration

Combining the two relations,

Energy = Mass x [Velocity]^2

Therefore, E/m = [Velocity]$^2 = c^2$

here c is unique velocity which is always constant.

$E/m = k_E/k_m = c^2$

Therefore, $k_E = hc$ and $k_m = h/c$, h here is universal constant

$4\pi r\nu = k_v$ = velocity = c

$E = k_E/4\pi r = hc/4\pi r = hc.(\nu/c) = h\nu$, so h is Planck's constant

The dimensional derivation of constants k_m and k_E

$4\pi rm = k_m$

the constant k_m has got dimensions of (Length).(Mass) = h/c

therefore, $k_m = h/c$

similarly, in the relation $4\pi rE = k_E$

k_E has got dimensions equal to Mass.(length)^3 / (Time)^2 = hc

therefore, energy relation becomes $4\pi rE = hc$

The other way of calculating the constants is given below:

In case of an Akashon having zero rest mass and moving with unique terminal velocity equal to c. If somehow the zero rest mass Akashon is brought to rest, the energy stored in the oscillating shells is radiated out in the form of waves. The source of waves are shells located in the region of perturbation. The velocity of the waves which originate from the shells must be equal to the velocity of Akashon because otherwise the continuity of wave trains is broken. It shall be shown later on also while discussing the dual nature, particle like and wave like, of the Akashon, that in case of the Akashon having zero rest mass and moving with unique velocity c, each shell is a source of wavelets moving with velocity c. the frequency of the wavelets which originate from the shells is given by principle of inverse variation. Suppose the shell is located at distance r from the geometric center,

$4\pi r \nu = k_\nu$

If λ is the wavelength of the wavelets

$\nu \lambda = c$

from the above two relations it is expected that

$\lambda = 4\pi r, \qquad k_\nu = c$

$4\pi r \, \nu = c$

The above relation can be considered as general in nature which means that it is applicable to all the Akashons with any mass whether moving or stationary and is also applicable to find out the frequency of the shells associated with the Prakashons.

$E = k"c / 4\pi r$

In case of Prakashon

$4\pi r \, \nu = c, \qquad => 4\pi r = c / \nu$

$E = k" \nu$

Therefore, the constant k" is equal to the traditional Planck's constant h.

After putting the values of the constants, the three basic equations based on the principle of inverse variation are now suitable for quantitative discussion.

1. $4\pi r\nu = c$ or $4\pi r.(1/t) = c$, where t is the time required to complete one oscillation. The frequency equation is used to find out the frequency of the independent oscillations of shells located in the region of perturbation. In case of Prakashon (photon) $4\pi r$ is equal to wavelength of radiation.
2. The mass equation
   $4\pi r \, m = h / c$
3. The energy equation
   $4\pi r \, E = hc$

In all the above equations the benchmark velocity c is equal to the velocity of light. Actually the invariant constant c was introduced out of mathematical necessity. The dimensions of the constant c are that of speed and in nature its values happen to be equal to the velocity of light. But the benchmark constant c is invariant. The constant c having dimensions of velocity is independent of frame velocity just like other universal constants such as h, G etc. The constant happens to have value equal to velocity of light.

DERIVATION OF FORCE RELATION:

If l is the distance of the separation of the mass center and the geometric center of the Akashon having radius 'r' of the sphere of tranquillity, the magnitude of the secondary force is got by using the displacement relation for the measurement of the force. Just like Hooke's Law of Force

$F = F_0.(l/r)$

The relation is similar to the Hooke's stress and strain relation. Due to displacement of mass center on account of impressed force, energy is released which is stored as mattenergy in the temporary region of perturbation. Hence as 'm' increases, the radius 'r' of sphere of tranquillity becomes less and less. At a certain stage radius of sphere of tranquillity becomes equal to l. This stage leads to the state of maximum impressed force because mass center cannot be located outside the sphere of tranquillity. The radius of sphere of tranquillity is also minimum, hence Akashon accumulates maximum mass. Each Akashon reaches state of maximum force and maximum mass. If the force acting on the Akashon is gradually increased,

When, $l = r$
$F_{max.} = F_0(l/r) = F_0$

It is thus seen that force factor F_0 is equal to maximum force.

Evolution of Physical Laws

Let us derive the the equation which gives value of F. As already discussed,

$l = 4at^2 / \pi^2$

$4\pi r_0 t_0 = 4\pi r t$

t_0 and t are the inertial time intervals of the stationary and the Akashon (mass packet) in motion.

$t = r_0 t_0 / r$

$\Rightarrow t^2 = t_0^2 \cdot r_0^2 / r^2$

also, $F = F_0 . l / r$

$\quad = (F_0 / r).(4 / \pi^2).at^2$

$\quad = \{(F_0 / r).(4 / \pi^2).a.t_0^2 r_0^2\} / r^2$

$= \{(4F_0 t_0^2 / \pi^2).(a / r.(r^2/r_0^2))\}$

put, $r / r_0 = B$ = dimensionless variable

and from $4\pi r m = k_m$, $r = k_m / 4\pi m$

therefore, $F = 16 (F_0 t_0^2 / \pi k_m).(ma / B^2)$

The above secondary force relation is general in nature. It is Newton Force Relation because it involves the acceleration of the Akashon (mass packet)

Let us suppose, $16 (F_0 t_0^2 / \pi k_m) =$ constant $= \eta$

therefore, $F = \eta . ma / B^2$.

η is not a universal constant, its magnitude depends on t_0 which is variable. It is seen that starting with fundamental Hooke's Force Relation involving stretching; the force relation is changed over to Newton Dynamic Force Relation involving acceleration. The above force relation is also called Newton's second law of motion.**The treatment adopted here shows that the second law is not empirical because it has got theoretical justification also.**The derivation of Newton's second law is based on Hooke's force law for elasticity therefore credit for it goes to Hooke who was the chief antagonist of Newton.Similarly it will be seen later on that the derivation of Newton's gravitational force is also based on Hooke's force relation.At last Hooke has won the battle.

B is dimensionless variable and its value change with velocity because the value of r of the moving Akashon is less than that of stationary Akashon. If the velocity of the Akashon is very less, the volume of the extra temporary region of perturbation created as a result of motion will be very very less and hence r is nearly equal to r_0, therefore, B = 1. For slow moving Akashons, the force relation can be written as given below

$F = \eta.ma$

When the Akashon (etheron) is moving, all the bounded shells move with it without experiencing any resistance. It carries with it all the bonded ether allotted to it. In other words Akashon is universe unto itself.

MEASUREMENT OF FORCE AND DERIVATION OF FORCE RELATION:

When the force is acting on the Akashon, there are two ways of measuring the force. The, magnitude of the displacement of the mass center and the geometric center is direct measure of the secondary force acting on it. The force has also been related to the acceleration of the Akashon, therefore, there are two ways of defining the unit force, either by fixing the displacement of the mass centre or specifying the acceleration produced in the known mass by the force. The first concept is called static or Hooke's concept of force and the second is Dynamic or Newton concept of force. **The Hooke's force is fundamental force because derivation of Newton force is based on it therefore both the forces are equivalent.**

The Hooke's force = $F_0.(l/r) = (m.a)/(B^2)$ = Newton force

According to the second concept, the unit force can be defined as that much force which produces unit acceleration when, $m/B^2 = 1$. According to the above concept of specifying the unit force

$F = 16.(F_0 t_0^2 /\pi k_m).(ma/B^2)$
$16.(F_0 t_0^2 /\pi k_m) = 1$
$\Rightarrow F = ma/B^2$

The above fore equation gives the force experienced by the Akashon in the defined force units.

If the displacement concept force is used, the unit force can be defined as $(\pi k_m/16 t_0^2)^{th}$ part of force factor. F_0, F is the value of force when mass center and geometric centers are separated by distance 1. From the above definition of unit force

$$F_0/\pi k_m/16 t_0^2 = 16(F_0 t_0^2/\pi k_m) = 1$$

by putting the above value in the general force relation, the force F in new units is given by,

$$F = ma/B^2$$

The final force relation in stretched units is the same as in dynamic force units, therefore, the two force units are unified.

It is seen that the value of F is the same whether the static unit (displacement) or the dynamic unit (acceleration) of the force is used to measure the magnitude of the force. The static and the dynamic concepts of the force are thus unified.

Let us next determine the value of F_{max}, the limiting maximum force experienced by the special stationary Akashon having radius r_n. Suppose the Akashon experiences the F_{max} force for the inertial time interval t_n specified for the special stationary Akashon. The mass center will shift away to just touch the surface of the sphere of tranquillity during the inertial interval t_n but the Akashon as a whole will not move because it is experiencing force for time interval just equal to the inertial time interval.

The maximum limiting acceleration $a = (\pi^2/4t_n^2).l$

When the acceleration is limiting acceleration $l = r_n$

$\Rightarrow a = (\pi^2/4t_n^2).r_n$

also, $r_n = k_m / 4\pi m_0$

therefore, $a = \pi k_m / 16t_n^2 m_0$

$\Rightarrow m_0.a = \pi k_m / 16t_n^2 = F_{max.} =$ Maximum Force

The maximum force F_0 or $F_{max.}$ can also be derived by changing F_0 to F_{max}

$16.(F_0 t_n^2/\pi k_m) = 1$, therefore

$F_0 = \pi k_m/16t_n^2$

$k_m = h/c$

$F_0 = \pi h/16ct_n^2$

Calculation of inertial time interval in achieving state of maximum force:

One way of calculating inertial time interval is by applying the principle of inverse variation $4\pi r_0 t_0 = k_t$. Alternative method of calculating inertial time interval is also there. It is klnown that when state of maximum force is reached the mass center just touches the sphere of tranquillity to merge with the center of oscillating shell in the surface of sphere of tranquillity. The center of oscillating shell is expected to be in state of mean position or zero force. The above condition is possible if during inertial time interval the shell just completes half oscillation.

Time required forcompletinghalfoscillations by applying the principle of inverse variation:

$4\pi r_n v_0 = c$

$4\pi r_n \cdot (1/t_n) = c$

Inertial time interval = time required for completing half oscillation = $2\pi r_n/c$

Put the value of t_n in the maximum force relation,

$F_{max} = \pi h/16 c t_n^2$

$t_n = 2\pi r_n/c$

therefore, $F_{max} = hc/64\pi r_n^2$

$4\pi F_{max} r_n^2 = hc/16$

Thus, for calculating maximum force relation, principle of inverse variation is applicable with the difference that r is replaced by r^2. It can be said that to calculate vector magnitude r is replaced by r^2.

For the Akashon already in state of motion it can be seen that the value of limiting force F_{max} is same constant value as deduced above and later on also. t is the inertial time interval for the Akashon (mass packet) in motion. From the above discussion about the nature of the force the following important conclusions can be drawn –

There is time gap between the application of the force and the change in the state of motion as a whole because during the inertial time interval the mass center is just shifted away from the geometric center. The inertial time interval of the stationary Akashon with zero rest mass is zero, therefore, such as Akashon will start moving as soon as it experiences force.

The Mass Energy Equivalence:

It has been discussed that the cause of birth of mass (Akashon) and energy (Prakashon) is the perturbation in the stretched ether (Akasha) and anti-ehter (anti-Akasha). The progenitor of mass and energy being same, therefore, mass and energy are equivalent.

According to the principle of inverse variation, the mass and energy are given by the following relations.

For mass packet, $4\pi r_m = k_m$

For energy packet, $4\pi r_E = k_E$

If mass packet having radius of sphere of tranquillity equal to r_m changes to energy packet having the same radius $r_m = r_E$, on dividing the two relations

$E/m = k_E/k_m = k$

$E = k.m$

k is equal to ratio of two proportionality constants hence it is universal or invariant in nature.

Radius of Spheer of Tranquillity of Akashon in Motion and derivation of γ:

The general force relation $F = ma / B^2$

The energy mass relation $E = km$.

In the above two equations B and k are involved. k is universal constant. Let us next find out the values of B and k.

The concept of energy which follows the principle of inverse variation has already been introduced. The energy is equal to the product of force and the distance moved along the point of application of force. Suppose the mass center of the Akashon (mass packet) moves through distance dS under the influence of the secondary force, therefore, the extra energy pumped into the Akashon is calculated as given below

$$dE = F.dS = (ma/B^2).dS = (m/B^2).(dv/dt).dS$$
$$\Rightarrow dE = (m/B^2).(dS/dt).dv = (mv/B^2).dv$$
$$B = r/r_0 = m_0/m$$

m and m_0 are the masses of the Akashons (mass packet) in the state of motion and rest.

$$dE = (m_0.v/B^3).dv$$

k is universal constant and invariant under all conditions
$$E = k.m$$
$$m = m_0/B$$

Again on differentiating
$$dm = -(m_0/B^2).dB$$
$$dE = k.dm = -k.(m_0/B^2).dB$$

Equating the values of dE got by the two methods
$$(m_0.v/B^3).dv = -(k.m_0/B^2).dB$$
$$v.dv = -kB.dB$$

The velocity v is measured relative to a point. Initial velocity relative to point is zero, and final relative velocity is v.

When the Akashon is at rest $v = 0$, $B = 1$

$$\int_0^v v.dv = -k \int_1^B B.dB$$

$$v^2/2 = -k/2.(B^2 - 1)$$

$$B = \pm \sqrt{\{1 - (v^2/k)\}}$$

Negative sign of B predicts the presence of anti-matter. B is dimensionless, therefore, the constant k should have dimensions of the square of velocity. The constant is universal and invariant. The invariant velocity happens to be equal to the velocity of light c. The above point is discussed later on also.

$$r = r_0 \sqrt{\{1 - (v^2/k)\}} = r_0 \sqrt{\{1 - (v^2/c^2)\}}$$

The above discussion shows that there is decrease in the radius of sphere of tranquillity of that Akashon only which has been subjected to force and hence has got real motion.

$B = r/r_0 = m_0/m$

$m_0/m = \sqrt{\{1 - (v^2/c^2)\}}$

$v^2 = k\{1 - (m_0^2/m^2)\}$

put $m_0 = 0$, $v = \sqrt{k}$ = unique constant velocity = c

The unique value happens to be equal to velocity of light.

The above result shows that the Akashon having zero rest mass can have the unique velocity $\sqrt{k}$ while for the Akashon having finite rest mass $v < \sqrt{k}$. The velocity c is measured relative to the same point as v is measured, hence velocity of light is constant relative to point of origin.

$$m = m_0 / B = m_0 / \sqrt{\{1 - (v^2/c^2)\}}$$

The above result shows that the mass of the moving Akashon is more than the Akashon at rest. The result is expected, because when the Akashon at rest is set into motion, the energy imparted to the Akashon is used in creating the temporary region of perturbation. The basic relations of special theory of relativity has been derived without using Lorentz'stransformation equation. The value of $1/B = 1/\sqrt{\{1 - (v^2/c^2)\}} = \gamma$ has been derived without considering frame velocities. c is here only a constant having dimensions of velocity while deriving γ no thought experiment involving light radiation was used. No concept of processes taking place simultaneously were used. Constant c was introduced out of mathematical necessity because it is proportionality constant, it is coincidental that its value was found to be equal to velocity of light. Actually it is benchmark velocity as discussed further.

$$r = r_0 \cdot [1/\sqrt{\{1 - (v^2/c^2)\}}]$$

The above equation shows that if $v > c$, r will be imaginary hence, speed limit of Akashon is c.

$$v = c[1/\sqrt{\{1 - (r^2/r_0^2)\}}]$$

If r_0 or rest mass of the Akashon is finite, it can be speeded upto any velocity less than c. If r_0 approaches infinity, $m_0 \rightarrow 0$ or rest mass of Akashon is zero.

v = c, hence Akashon having zero rest mass can be speeded upto one and only one unique velocity c relative to the starting point wether the starting point is in motion or at rest w.r.t. the observer. The photon is said to have zero rest mass **hence velocity of photon is always equal to c relative to the source.** The above discussion proves the invariance of speed of light. The velocity of light has to be constant w.r.t. source of light due to mathematical necessity because velocity of light happens to be equal to square root of constant k which is equal to ratio of two constants of proportionality. Einstein arbitrarily assigned the property of invariance to light velocity but here reason is there because the magnitude of velocity happens to be equal to square root of a universal constant having dimensions of velocity.

Radius of the sphere of tranquillity when Akashon is subjected to acceleration:

It is seen that when the Akashon is moving with uniform velocity v,

$$r = r_0 \cdot [1/\sqrt{\{1 - (v^2/c^2)\}}]$$

If the Akashon is moving towards a point and is subjected a constant acceleration a, after time interval t,

$$r = r_0 \cdot [1/\sqrt{\{1 - (a^2 t^2/c^2)\}}]$$

Evolution of Physical Laws

The case of variable acceleration:

Suppose the test Akashon is attracted by central force, is the acceleration of the Akashon depends on the distance r of the test Akashon from the point at which the central force is situated. Suppose acceleration
$a = \alpha/R^n$, where α and n are constants.

Force F acting on test Akashon, $F = ma/B^2 = (m.\alpha)/(B^2 R^n)$,
m is the mass of the Akashon and R is the distance of Akashon from the central force point.

Change in energy dE as test Akashon moves through distance dR,
$dE = -(m.\alpha/B^2).(dR/R^n)$
$m = m_0/B$, $dE = -(m_0.\alpha/B^3).(dR/R^n)$

As already seen, in case of uniform speed
$dE = -k(m_0/B^2).dB$

Equating two values of energy,

$$-\alpha \int_{\infty}^{R} (dR/R^n) = -k \int_{1}^{B} B.dB$$

When Akashon is at rest $R \to \infty$, $B = 1$

$B = \sqrt{[1 - [2\alpha/\{(n-1).k.R^{n-1}\}]]}$

$B = r/r_0$

$r = r_0.[\sqrt{[1 - [2\alpha/\{(n-1).k.R^{n-1}\}]]}]$

If the central force of gravity and the mass of the gravitating body is M, $\alpha = GM$, $n = 2$, k as already discussed $= c^2$, therefore when the test Akashon is at a distance R from the gravitating body,

$$r = r_0 \cdot [\sqrt{\{1 - (2GM/c^2R)\}}]$$

In the above equation, r_0 is the radius of sphere of tranquillity when the Akashon is at infinite distance or no gravitational force is there, r is the radius of sphere of tranquillity when the Akashon is in gravitational field. The above equation shows that there is length contraction in the gravitational field.

The above relation can be deduced more easily also,
Kinetic Energy = Gravitational Potential Energy
$\frac{1}{2}(mv^2) = G \cdot m \cdot M/R$
$v^2 = 2GM/R$

Putting the value of v^2 in,

$$r = r_0 \cdot [\sqrt{\{1 - (v^2/c^2)\}}]$$

$$r = r_0 \cdot [\sqrt{\{1 - (2GM/c^2R)\}}$$

The above equation in the gravitational field will be used to discuss general theory of relativity, while equation $r = r_0 \cdot [1/\sqrt{\{1 - (v^2/c^2)\}}]$ is the backbone of special theory of relativity as will be discussed later on.

Expression for Momentum:

Momentum $p = m.v$

$m = h/4\pi rc$

$v = c\sqrt{\{1 - (r^2/r_0^2)\}}$

v is got from the relation $r = r_0\sqrt{\{1 - (v^2/c^2)\}}$

Putting the values of m and v,

$p = (h/4\pi r).\sqrt{\{1 - (r^2/r_0^2)\}}$

$4\pi r.p = h.\sqrt{\{1 - (r^2/r_0^2)\}}$

So long as r does not change (state of motion) $4\pi rp$ remains constant. Therefore the principle of inverse variation is applicable to p or momentum is fundamental property.

For calculating momentum according to above expression, measurement of only r is required provided r_0 is known. From the usual expression, $p = mv$ measurement of mass and velocity is required. Further, velocity measurement require measurement of length and time, therefore, three measurements are needed. While in the changed expression, measurement of only r is required.

Rate of Change of Momentum with Time:

$dp/dt = (dp/dr).(dr/dt)$

$p = (h/4\pi r).\sqrt{\{1 - (r^2/r_0^2)\}}$

$dp/dr = -(h/4\pi).\{1/(r^2\sqrt{1 - (r^2/r_0^2)}\} = -(h/4\pi)(1/r^2)(c/v)$

$$r = r_0\sqrt{1 - v^2/c^2}$$

$dr/dt = -(v/c^2).a.[r_0/\sqrt{\{1-(v^2/c^2)\}}] = -(v/c^2).a.(r_0^2/r)$

a is acceleration

$dp/dt = (dp/dr).(dr/dt)$

Putting the appropriate values,

$dp/dt = (ha/4\pi r_0 c).\{1-(v^2/c^2)\}^{3/2} = ma/\{1-(v^2/c^2)\} = $ Force

Quantization of mass and energy: The mass of akashon = $h/4\pi rc$ and energy of prakashon = $hc/4\pi r$ It is known that minimum measurable length a_0 is of the order of 10^{-35}m. Suppose there are two values of radii of spheres of tranquillity r and r_0 of moving akashon at at rest, $r=na_0$ and $r_0 = n_0 r_0$. The growth of the akashon mass or that of mass packet is inwards contrary to the conventional growth which is outwards. Mass difference of two akashon=$h/4\pi ca_0$, $(n-n_0)=N=1,2,3,4,...$ Therefore, growth in mass=$h/4\pi ca_0\{(N(n-n_0)/nn_0)\}$. It shows that growth in mass is not continuous it is quantized. Similarly it can be shown that growth of energy of photon is also quantized. Similarly it can be shown that growth or difference in energy of two photon=$hc/4\pi a_0\{(N(n-n_0)/nn_0\}$ it shows that change in energy of photon is not continuous it is quantized.

Quantization of velocity:

The relation $r = r_0\sqrt{1-v^2}/c^2$ has been proved in the last topic.
From it $v^2 = c^2(1-r^2/r_0^2)$ r = na_0 and $r_0 = n_0 a_0$, the symbols have got usual meaning.

$$v^2 = c^2 \{(n_o + n)(n_o - n)\}/n_0^2 n_0 \quad -n = N$$
$$N = 1, 2, 3, \ldots \ldots v^2 = c^2 \{N(2n_0 - N)\}/n_0^2.$$

The relation shows that square of velocity is quantized. For lighter akashon n_0 is very huge, for electron it is $= 1.67 \times 10^{22}$. Therefore, $2n_0 \gg N$
$$v^2 = 2Nc^2/n_0$$

It shows that increase in velocity is not continuous, it increases in stages. The increase in velocity at each stage is $2c^2/n_0$ this is also the minimum value of velocity which can be imparted to light akashon. No lighter akashon can move with velocity less than value or this is the mandatory natural value of velocity.

Terminal velocity: the terminal velocity has also been discussed afterwards $v^2 = c^2(1 - n^2/n_0^2)$, $r = na_0$ r cannot be less than the minimum value of natural length, hence minimum value of $n=1$ the square of velocity equation shows that as n decrease s the value of velocity goes on increasing. The maximum value of velocity achieved is when $n=1$, $r=n.a_0$. Therefore, at this stage the radius of sphere of tranquillity r of moving akashon is reduced to minimum value a_0 at this stage no further decrease in radius is allowed. The maximum value of velocity attained is called terminal velocity.

The terminal velocity $v_c^2 = c^2((1 - 1/n_0^2))$
$$v_c^2 = c^2(1 - a_0^2/r_0^2),$$
r_o is the radius of sphere of tranquillity of akashon at rest.

As the radius of tranquillity $r_0 = h/4\pi m_0 c$ depends on mass at rest hence, the value terminal velocity is related to mass at rest. At the stage of terminal velocity the radius of sphere of tranquillity of moving akashon is reduced to minimum value a_0 hence the mass attained by moving akashon is maximum. The akashon with the maximum mass is named as N AKASHON. If the N akashon moving with terminal velocity is speeded up further it will disintegrate into small fragments.

The mass of N AKASHON $M_N = h/4\pi c a_0$ the value of mass is deduced later on also. In case N akashon $n_0 = 1$ hence terminal velocity of N akashon is zero which shows that it has got unique property that it cannot me moved however great the force is applied. It is in state of permanent rest. The unique N akashon disintegrates if it starts moving.

Quantization of kinetic energy:

Suppose kinetic energy is calculated by using relativistic mass –

Kinetic energy = $1/2 mv^2$, $v^2 = c^2(1 - r^2/r_0^2)$ $r = n a_0 r_0 = n_0 a_0$

$r/r_0 = m_0/m = n_0/n$

Using the above relations,

kinetic energy = $(1/2)mc^2(n_0^2 - n^2)/n_0^2$ = $1/2 m\, c^2\{((n_0+n)(n_0-n))\}/n_0^2 = \{(n_0+n)/2n_0\}(\Delta m)c^2$

here, Δm is the increase in mass as result of motion hence second factor is energy equivalent of mass increase.

If speed is slow $n_0 \approx n$ hence first factor is equal to one which leads to the conclusion that kinetic energy is equal to energy equivalent of mass increase.

If speed is very high $n_0 >> n$ which shows that $2n_0 >> (n_0 + n)$. The first factor in the above expression is less than one therefore kinectic energy is less than increases in mass equivalent energy.

Mass equivalent energy = kinectic energy + inertial mattenergon energy.

Kinetic energy calculated here is the relativistic kinetic energy because relativistic mass is used inertial mattenergon is created during inertial time interval when the akashon resists change in state while relativistic mass is created after the akashon starts moving .Inertial mattenergon energy can also be named as zero point energy. The inertial mattenergon energy is stored during inertial time interval when the akashon is set into motion. The moving akashon comes to conventional rest during inertial time interval, the packed inertial mattenergon energy is dissipated or used up to overcome the resistence force when the moving akashon comes to rest during inertial time interval needed to bring the moving akashon to rest. The inertial mattenergon or zero point energy is thus useless energy because it is created when akashon resists change in state of rest to motion and is consumed when akashon resists change in state of motion to rest. Therefore, inertial mattenergon energy or zero point can also be named as resistance energy. It is thus seen that energy start packing up as inertial mattengon during inertial time when mass packet is not moving, kinetic energy or execss mass energy appear later on when particle starts moving.

Kinetic energy=1/2mv^2 putting value of velocity Knetic energy=Nmc^2/n$_0^2$
It is seen kinetic energy is quantized because N=1, 2, 3, ……….

THE COSMIC TERMINAL VELOCITY OF ELEMENTARY PARTICLE:

When an Akashon is subjected to secondary force, the mass center starts moving away from the geometric center and during this shift energy is liberated which is used in creating temporary region of perturbation or extra mass. The increase in massas a result of motion is given by the following relation as already discussed

$$m = m_0 / \sqrt{\{1 - (v^2/c^2)\}}$$

The above equation shows that m →∞ as v → c, but practically it is not possible because each elementary particle cannot be speeded up beyond a certain velocity called terminal velocity, hence there is limit to the relativistic mass.

As a result of increase in mass, surface of sphere of tranquillity moves inward and due to application of force mass center moves away from the geometric center towards the surface of sphere of tranquillity. At a certain stage of force applied or velocity the mass center will just touch the surface of sphere of tranquillity and subsequently no shift is possible because mass center can not be located outside the sphere of tranquillity. **The maximum velocity acquired by the Akashon at this stage is called cosmic terminal velocity.** The force applied at this stage is called maximum force. The mass accumulated by the Akashon at this stage is called maximum mass. It can also be said that when the Akashon is subject to outside force as a result of motion the mass of the Akashon becomes equal to limiting mass associated with the heaviest Akashon named as N Akashon or M_N. When the moving Akashon acquires the maximum mass M_N, no further change in velocity is possible. The velocity at this stage is called terminal velocity.

The mass of the N Akashon M_N as calculated later on = $\sqrt{(hc/4\pi G)}$ = 1.539×10^{-8} kg.

M_N Akashon undergoes metamorphosis if the velocity is increased beyond terminal velocity. M_N Akashon may breakup into photons and other particles having mass less than M_N.

In the mass velocity equation, put $m = M_N$ and $v = v_c$

$M_N = m_0 / \sqrt{\{1 - (v_c^2/c^2)\}}$

Cosmic terminal velocity $v_c = c\sqrt{\{1-(m_0/M_N)^2\}} = c\sqrt{\{1-(4\pi G m_0^2/hc)\}}$

The above expression shows that each Akashon has got terminal velocity which depends on rest mass m_0. The Akashon with zero rest mass has got maximum terminal velocity $v_c = c$.

Put value of M_N

$V_c = c\sqrt{\{1 - (4.22186 \times 10^{15} \times m_0^2)\}}$

In case of electron, putting the value of mass of electron

$$V_c = c\sqrt{\{1 - (3.5038 \times 10^{-45})\}} = (0.999999999999999999999999999999)c$$

Similiarly, by putting value of mass of proton in the general equation, the terminal velocity of proton $v_c = c\sqrt{\{1 - (1.0867 \times 10^{-19})\}}$

terminal velocity of proton $v_c = (0.999999999999999999)c$

The terminal velocity of proton is less than that of electron and both the terminal velocities are nearly equal to velocity of light.

So v_c is very very nearly equal to c in case of electron and proton. When the terminal velocity of electron or proton is reached the particles can not be accelerated further even when large force is applied. The above experiment related to the observation of terminal velocity can be performed in high energy cyclotron, which will conclusively prove **that any hadron can not be speeded up indefinitely.**

The terminal velocity of electron is nearly equal to the velocity of light. Similarly, terminal velocity of proton can also be calculated. The experiment to know the terminal velocity of a particle can be easily set up by subjecting the particle to force. At a certain stage there will be no change in the velocity when terminal velocity is acquired even when the particle is subjected to further force. At this stage, the particle may undergo metamorphosis or disintegrate when the Akashon attains terminal velocity, the mass acquired at this stage as a result of motion is equal to the N Akashon. The stage of terminal velocity can be easily witnessed by

Evolution of Physical Laws

allowing the charged particles to move in vacuum tube and subjecting the particles to electric field or any other device to produce force.

The Benchmark Terminal Velocity c:

From the above relation it is seen that if the rest mass of the Akashon $m_0 = 0$, the value of terminal velocity $v_c = \sqrt{k} = c$, the benchmark velocity. The Akashon having finite rest mass center cannot have the terminal velocity equal to c because the mass touches the sphere of tranquillity before the value of terminal velocity equal to c is attained.

The further increase in velocity is not then possible. The Aakashons having very very small value of a rest mass m_0 can attain the velocity nearly equal to c, for the Akashons having large value of rest mass; the value of terminal velocity is very less. Each Akashon has got terminal velocity, which depends on mass.

The terminal velocity relation is strictly applicable to true Akashon. When the cosmic terminal velocity is reached, the velocity does not change even with the further force applied. At this stage, the Akashon is metamorphosized. The limit of the mass of the Akashon up to which terminal velocity is applicable can also be determined. In the general relation for the terminal velocity, the factor under the square root must be positive. The factor remains positive up to the mass limit given below:

$1 - (m_0/M_N) = 0$

or, $m_0 = M_N$

Frame Velocity and c:

Suppose there are two systems of frames A and B. the frame B is moving with the velocity v relative to the frame A, suppose Akashon with rest mass m_0 is subjected to acceleration in the two frames. The value of terminal velocity in frame A is v_c and in frame B it is v_c'.

$v_c' = v_c + v$

$1 - (v_c^2/c^2) = (m_0/M_N)^2$

$1 - (v_c'^2/c^2) = (m_0/M_N)^2$

If the Akashon has zero rest mass $m_0 = 0$

$v_c' = v_c = c$

Therefore, the Cosmic terminal velocity of the mass packet having zero rest mass is unique velocity which is independent of the frame velocity. In other words the Akashon having finite mass can be used to calculate the relative velocity of the two frames after knowing the terminal velocities in the two frames but Akashon with zero rest mass is not suitable to determine the relative velocity of two frames. The universal constant can be put equal to c. Nothing can be added to or subtracted from the velocity c and that is why **Michelson Morley experiment gives negative results because here photon with zero rest mass is used to determine the relative velocity of earth w.r.t. ether taken as stationary.**

BENCHMARK VELOCITY AND CHANGE IN VELOCITY OF LIGHT IN VACUUM:

Actually only Akashon or photon having rest mass equal to zero can exactly achieve terminal velocity value c. **The velocity c is only ideal or benchmark velocity. Nothing in the universe can exactly attain this velocity.** Let us see why it is so. To have rest mass equal to absolute zero is singularity like infinite mass. The rest mass of photon is very very nearly equal to zero, therefore even the velocity of the photon is not exactly equal to c.

The radius of sphere of tranquillity of Akashon or photon having zero rest mass is infinite. The inertial time interval is given by the following equation:

$4\pi rt = k_t$

$t = k_t/4\pi r$

when $r \to \infty$, $t \to 0$

but both the above two results lead to singularity.

It means that zero time interval is required for mass packet with zero rest mass to start moving from rest. Zero time interval is singularity which is against the nature of things because any process taking place in nature requires some finite time. Apart form zero inertial time it can also be seen that Akashon (Photon) with zero rest mass will never move because though no time is needed for Akashon to start moving but total time required to exactly achieve terminal velocity c will be infinite because the rest value of radius of sphere of tranquillity is infinite and when the Akashon acquires terminal velocity c, the radius of sphere of tranquillity must drop from infinite value to some definite value, and this process will need infinite time or process cannot be completed. It can be said that the Akashon (Photon) with zero rest mass will never move. Infinite radius of sphere of tranquillity is again singularity. It can be said that nothing exist in the universe with infinite radius of sphere of tranquillity hence zero rest mass. It is applicable to the photons and neutrino also. The rest mass value of photon is not exactly zero though it is nearly zero. It can be said that the different photons have different rest mass values, which are nearly equal to zero.

Suppose photon starts with rest mass m_0 and m is the mass (energy) when it acquires terminal velocity.

$$m = m_0 / \sqrt{\{1 - (v_c^2/c^2)\}}$$

the terminal velocity v_c is given by the following relation as already deduced:

$$v_c^2 = c^2 \{1 - (4.22186 \times 10^{15} m_0^2)\}$$

The photon rest mass m_0 is not exactly zero. Therefore, $v_c < c$ or it can be said that the speed of photon is less than benchmark velocity c.

From terminal velocity relation,

Evolution of Physical Laws

$$m = m_0 / \sqrt{\{1 - (v_c^2/c^2)\}} = F_{max}/a$$

From the terminal velocity relation, it is seen that photons having more rest mass will have velocity less than the benchmark velocity c as compared to the photons having less rest mass. Also the mass of the moving photon will be more if rest mass is more.

The photon starting with more zero rest mass will have more mass on achieving terminal velocity as compared to photon with less zero rest mass. From cosmic terminal velocity equation it has been observed that photon with more rest mass will have less terminal velocity which leads to the final conclusion that energetic photon move with less velocity as compared to photon having less energy and hence velocity of light is not conastant, it is frequency dependant. The difference in the velocities of photon is very very less or photon velocity is nearly equal to benchmark velocity c. It is only in case of photons coming from deep space after travelling huge distance that frequency or colour dependence of photon velocity can be detected. The change in vaccum velocity can also be qualitatively explained as given below.

According to special theory of relativity, the speed of light is taken to be constant in vacuum but it will be seen here that velocity of light is not the same for all photons but it depends on the energy of photons although the change in velocity is so small that it cannot be observed under ordinary conditions.

The frequency dependant velocity of light can also be discussed from new angle as given below. As already seen, according to the principle of inverse variation, the velocity of photon is given by the following relation i.e. $4\pi r\nu = c$, ν is the frequency of oscillation of shells located at distance r from the center. The shell density increases with decrease in value of r. With decrease in the value of r, ν also increases according to the principle of inverse variation but due to over crowding of shells, the frequency is reduced more than the value given by the principle of inverse variation which leads to the conclusion that electromagnetic radiation which starts from each shell is not constant equal to c, but it depends on the location of shell in the region of perturbation therefore wave packet is formed. The velocity of wave packet is less than c. The photons having higher value of ν are most affected because of dense packing of shells which leads to the conclusion that the velocity of energetic photons is less than the usual velocity c. The change in velocity is so less that it can be observed only if two photons having different energies reach a point after traveling billions of kilometers and then more energetic photon will be detected later on by the detector.

The rest masses (energies) of all the photons though very very small, but are not same, which means that energetic photons in motion will have more rest mass. If the value of rest mass or energy is put in the terminal relation it will be seen that terminal velocity of energetic photon will be less than as compared to the photon having less energy. it can be said that velocity of light depends on colour ($h\nu$) because energy depends on frequency. Suppose, c is the terminal velocity of photon if rest mass is exactly zero and

v_c is the actual terminal velocity, putting the value of these velocities in the terminal velocity, rest mass of photon can be determined. The same behaviour is also applicable to neutrinos also because rest mass of neutrino is not equal to absolute zero.

EQUIVALENCE OF MASS AND ENERGY:

Let us next see how much energy is pumped into the Akashon at rest or in uniform motion when it is subjected to external force.

$$dE = F.dS = F.(dS / dt).(dt / dv).dv$$
$$dE = F.(v / a).dv$$
$$F = ma / B^2 = m_0 a / (1 - (v^2 / c^2))^{3/2}$$
$$\int_0^E dE = m_0 \int_0^v [v / \{1 - (v^2 / c^2)\}^{3/2}].dv$$

on integration, $E = m_0.c^2 / \sqrt{\{1 - (v^2 / c^2)\}} - m_0.c^2$

$$\Rightarrow E = m.c^2 - m_0.c^2 = (m - m_0).c^2 = \Delta m.c^2$$

Δm is the temporary mass stored in the temporary region of perturbation created as a result of motion caused by the force. The region of perturbation created when the shells are squeezed inwards by the action of primary Cosmic Force is permanent but the region of perturbation created as a result of secondary force is manifested as temporary region of perturbation. When the Akashon moving with very high velocity is suddenly stopped, the temporary region of perturbation vanishes and the energy stored in it is carried away in some manner such as by the gravitational waves or by the gravitons.

Mass energy equivalence results can also be achieved from the first principle. Suppose the Akashon (mass packet) is subjected to force F, which brings about separation l in the mass center and the geometric center during the inertial time interval t.

$F = F_0 \cdot (1/r)$, $dE = F \cdot dl = F_0 \cdot (1/r) \cdot dl$

Acceleration, $a = k' \cdot l$

$dv/dt = k' \cdot l$, $\qquad (dv/dl)(dl/dt) = k' \cdot l$

$dl/dt = v$, $\qquad v \cdot (dv/dl) = k' \cdot l$

$v \cdot dv = a \cdot dl \qquad dl = (v/a) \cdot dv$

inertial time, $t = t_0 / \sqrt{\{1-(v^2/c^2)\}}$

$r = r_0 \sqrt{\{1-(v^2/c^2)\}}$

$l = (4/\pi^2) \cdot at^2$

$dE = (4F_0 at^2 / \pi^2 r) \cdot dl$

$dE = [4F_0 at_0^2 / \pi^2 \{1-(v^2/c^2)\}] \cdot [1/r_0 \cdot \sqrt{\{1-(v^2/c^2)\}} \cdot (v \cdot dv/a)]$

$\Rightarrow dE = (4F_0 t_0^2 / \pi^2) \cdot (1/r_0) \cdot [1/\{1-(v^2/c^2)\}^{3/2}] \cdot v \, dv$

$r_0 = k_m / 4\pi m_0$

$dE = 16(F_0 t_0^2 / \pi k_m) \cdot [m_0 / \{1-(v^2/c^2)\}^{3/2}] \cdot v \, dv$

As already seen $F_0 = \pi k_m / 16 t_0^2$

Therefore,

$\Delta E = m_0 \cdot \int_0^v [v/\{1-(v^2/c^2)\}^{3/2}] \cdot dv = [m_0 \cdot c^2 / \{1-(v^2/c^2)\}^{1/2}] - m_0 \cdot c^2$

$= mc^2 - m_0 c^2 = (m - m_0) \cdot c^2$

$= \Delta m \cdot c^2$

If $m_0 = 0$, $E = mc^2$

SIZE OF PHOTON:

The region of perturbation of photon extends upto infinity therefore the exact size of photon is limitless but the most of the energy of photon is concentrated in the region near the surface of sphere of tranquillity therefore, the effective size of photon is equal to the diameter of sphere of tranquillity of photon. In the above discussion it has been seen that radius and wavelength are related.

$4.\pi.r = \lambda$

$2.\pi.d = \lambda$

the effective diameter will be more than d

therefore, $d > \lambda/2\pi$

Cosmic force and refraction:

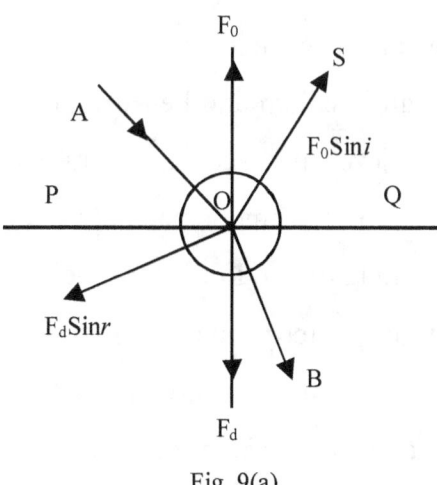

Fig. 9(a)

It will be seen that refraction of light proves the presence of cosmic force. The anti akasha shells of photon are equally stretched in outwards direction and no resultant force acts on isolated photon as already explained. The cosmic force acting in direction perpendicular to the direction of motion of photon follows the principle of inverse variation.

$4\pi rF =$ Constant

It is just like the law of friction or here space friction.

Suppose photon is about to enter the denser medium along AO direction, PQ is the surface of dense medium, F_0 is the resultant of cosmic force of the upper hemisphere of photon which is about to enter the dense medium acting along OC F_d is the resultant of cosmic force acting along OE in the lower hemisphere of photon which lies in the denser medium .It is obvious that $F_0 < F_d$

Therefore the photon which is about to enter the denser medium along path AO is pushed inwards due to cosmic force being more in the denser medium as compared to rare medium and hence follows the new bent path OB in the dense medium.it is thus seen that cause of refraction is the unbalanced cosmic force. The component of force acting along OS perpendicular to the direction of motion of the hemisphere of sphere of tranquillity of that part of photon which is in rare medium is $F_0 \sin i$. Similarily the component of acting in directionOR perpendicular to the direction of motion of that part which is in the dense medium is $F_d \sin r$. The radius of sphere of sphere of tranquillity changes only when the photon is completely immersed in the dense medium. In the transition state when the photon is just to enter the dense medium the radius of sphere of tranquillity is equal to the value in the rare medium. Applying the principle of inverse variation

$4\pi r F_0 \sin i = 4\pi r F_d \sin r \quad \sin i/\sin r = F_d/F_r = n \quad n>1$

The above is the proof of Snells law based on cosmic force or the refraction proves the existence of cosmic force. When the the photon is completely immersed in dense medium the radius of sphere of tranquillity changes from r_0 to r_d and cosmic force changes from F_0 to F_d.

Applying the principle of inverse variation $4\pi r_0 F_0 = 4\pi r_d F_d$ using the relation $4\pi r$=wave length. There is no change in wave frequency $F_d/F_0 = \lambda_0/\lambda_d = c/v = n$ v is the value of velocity in dense medium. The above explanation of refraction also explains that when the photon renters the rare medium it regains its original values because the vacuum cosmic force is restored.

PROPERTIES OF AKASHONS AND PRAKASHONS:

Some very interesting results can be deduced from the basic equation by using simple calculations. The mass and the energy are not uniformly distributed in the boundless region of perturbation around the sphere of tranquility. The mass or energy density or shell density can be easily determined at any point in the region of perturbation. It is more apt to call it as density of morphosed akasha:

$4\pi r = h/c$

The differentiation leads to the following results. Neglecting negative sign

$dm/dr = h/4\pi r^2 \cdot c$

The mass dm is contained in the volume $4\pi r^2 \cdot dr$ of the region of perturbation. The mass density at distance r from the region of perturbation is given by the following equations neglecting negative sign

Mass density $\rho_m = dm/4\pi r^2 \cdot dr = (h/16\pi^2 \cdot c)\cdot(1/r^4)$

Putting the value of constants

$\rho_m = (1.399 \times 10^{-44}/r^4)$ kg.m^{-3}

the value of constant is very very small and there is inverse relation between the density and the fourth power of distance which results in very steep fall in the density with increase in distance.

The magnitude of density at various points can be calculated by assigning arbitrary values to the distance r

Suppose $r = 10^{-14}$ m

$\rho_m = 1.399 \times 10^{12}$ kg.m^{-3}

The huge value indeed

When $r = 10^{-10}$ m

$\rho_m = 1.399 \times 10^{-4}$ kg.m^{-3}

If $r = 10^{-9}$ m

$\rho_m = 1.399 \times 10^{-8}$ kg.m^{-3}

If $r = 1$ m

$\rho_m = 1.399 \times 10^{-44}$ kg.m^{-3}

The above value of density is very very small. It is thus seen that the density in the region of perturbation can change from very huge value to insignificant value. The density expressed here is not conventional density, it is shell densiy of Akasha or ether shells. It signifies the strength of influence or effective presence of the Akashon although the influence extends upto infinity, but it tapers off rapidly with distance. The Akashon may behave like a point mass if the mass is supposed to be concentrated at the centre of sphere of tranquillity but actually we have seen that it has got extended dimensions also. There is no conflict in the quantum theory and general theory of relativity, because general theory of relativity is based on point mass and quantum mechanics is based on probability which is possible only in extended mass. The both qualities are present in the model of Akashon presented here.

Size of Electron and Proton:

As seen above there is no limit to the size of photon and electron but effective size of electron and proton will be a little greater than the radius of the sphere of tranquillity. In case of pure mass packet there is only one radius of sphere of tranquillity, but in case of Akashon composed of hybrid charge mass shells like electron, proton there are two radii, one is associated with mass called mass radius, and the second is associated with charge is called charge radius. It is the charge radius of sphere of tranquillity that is experimentally measured.

Mass Radius of electron = $h/4\pi m_e.c$

Putting the values of constants,

Mass Radius of electron = 1.93×10^{-13} m

Similarly, the radius of proton = 1.05×10^{-16} m

Actually, influence of region of perturbation extend up to infinity but, it has been just seen that the mass density falls rapidly with increase in distance and practical influence extends up to the point where density is sufficient. The above size of electron or proton shows that these particles are not point like as has been hitherto assumed. The extended size of the Akashon or Prakashon can explain the dual nature of these particles. The extended mass can explain other properties also which will be discussed later on.

Entanglement:

The region of perturbation of the Akashon is boundless, therefore, the presence of Akashon should be felt up to infinity but because nearly whole of the mass of the Akashon is confined to extremely small volume around the sphere of tranquillity while the mass in the rest of the region is distributed so sparsely that its existence can be ignored, that is why the Akashon (mass packet) is supposed to have no sharp but fuzzy boundary. The pen in my hand has got no sharp boundary, its existence extends up to infinity. All the big and the tiniest of the things created by God are endowed with boundless dimensions. Everything being limitless, nothing is isolated in the universe. It is in the fitness of the order that the manifested tiniest of the creation of the Infinite creator should have infinite dimensions. There is interaction between each and everything however great the distance of separation may be. The human body has also infinite dimensions. Nothing is isolated in the Universe even the fall of a leaf disturbs the whole universe. The unification is expressed in *Hindu* scriptures, *Vedas* like this "that are thou". The same idea is expressed in *Mandyyuka Raraita Upnishada***"all objects are in origin unlimited like space. And multiplicity has no place in them in any sense"**. The Mystic poet Francis Thompson wrote –

"All things by immortal power near or far hiddenly to each other linked are that thou cannot stir a flower without troubling a star"

All the elementary particles are entangled because region of perturbation of each elementary particle is unlimited. The region of perturbation can interact even if the particles are separated by unimaginable distance;**therefore the spooky action at a distance is possible.**

The Infinite director is directing the perpetual Cosmic show in which the infinite actors participate on the infinite stage. Out of the unmanifested Infinite, the manifested Infinite has precipitated. As seen earlier, in case of Akashon (mass packet) nearly whole mass of the Akashon is present inside sphere having radius of the order of 10^{-9} m. because outside this region mass density of the region of perturbation is extremely small, therefore, huge amount of mass can be concentrated in tiny volume $(4/3)\pi r^3 = (4/3).\pi.(10^{-9}) = 4.19 \times 10^{-27} m^3$, provided the radius of the sphere of tranquillity is very less as compared to 10^{-9}m. In the volume equation 10^{-9}m is the rough estimate of the distance from the geometric center up to the point beyond which the mass density in the region of perturbation is so less that it can be neglected. By making the radius of the sphere of tranquillity less and less, huge amount of mass can be packed in the extremely tiny volume. When the radius of the sphere of tranquillity is equal to 10^{-75}m the mass packed in nearly $4.19 \times 10^{-27} m^3$ is equal to 1.8×10^{30}kg, the mass nearly equals to the mass of the sun. As the radius of the sphere of tranquillity becomes still smaller, very huge mass can be packed in the tiny volume $4.19 \times 10^{-27} m^3$. Theoretically according to the mass equation when r→0, m→∞, but practically it is impossible for the radius of sphere of tranquillity to be zero because of interaction of the shells present on the surface of the sphere of tranquillity which puts limit to the minimum value of the radius, but at the time of the origin of creation the shell radius might be very very small than the supposed present value of 1.14×10^{-35}m which leads to very very small value of radius of sphere of tranquillity, hence accumulation of huge amount of mass in very small volume but mass cannot approach infinite value because radius r cannot approach zero. The lowest limit of the radius of the sphere of tranquillity is very very less which leads to accumulation of unbelievable mass in the tiny volume which can be taken nearly equal to zero. The huge amount of mass (energy) concentrated in the nearly point volume of space exploded to be strewn all around with big bang like the pellets stuffed in an explosive device. The basic equation related to big bang theory is mass equation

$m = h/4\pi r.c$, or

the energy equation $E = hc/4\pi r$

It appeals to the mind that the mass might not have been concentrated at the single nearly point volume, but there might be many other point volumes fit for the big bang to happen. Moreover, there might be still other poin volumes though very far away which are still in the gestation process for the big bang to occur. It is most likely that the creation is continuous process. In the expanding Universe the stars are moving away after big bang but our Universe is limited up to the far sight limit of the eye of the telescope. There might be other point volumes, though very very far away where matter might be moving towards us after the big bang but the risk of collision can be rated as very slim in the near future. The Universe will get more crowded and chaotic with the passage of time and hence prone to cosmic accidents. The black hole might be single Akashon associated with huge mass as a result of very very small radius of the sphere of tranquility. The black holes are massive Akashons which are born as such because of very small radius of the sphere of tranquillity. The black holes are the Akashons which failed to explode. The black holes are not freak Akashons.

The freedom of the will of the God is expressed in the creation of tiniest of the neutrino and the massive black holes. The black holes are also associated with the sphere of tranquillity and the region of perturbation is much dense as compared to the lighter Akashons. The lighter Akashons can be easily assimilated by the black holes to become part of the region of perturbation of the black holes. The black holes can be called elementary particles with huge mass. Black hole at the center is just like elementary particle having mass $m = h/(4\pi rc)$ but in case of black hole, $r \to 0$ because it is composed of shells having radius nearly equal to zero. In case of lighter elementary particles the value of r_0 is much larger as compared to shells of black holes. The actual size of the black hole is nearly a point but it has got large apparent size because central mass point is surrounded by ordinary matter in very dense state which is to be devouvred by the central mass.

THE BIG BANG AND BLACK HOLES:

There are four types of shells two of akasha $A(+)$; $A(-)$ and two of anti akasha $A^{\wedge}(+)A^{\wedge}(-)$. The four shells combine to form tetrad of shells. Tetrad shells are used to form BRAHAMAND (cosmic egg). The shell radius of tetrad is much smaller than that of individual shells hence limiting radius of sphere of sphere of tranquillity of Brahamand is nearly zero which leads to the conclusion that Brahamand is composed of nearly infinite number of shells. The tetrad shells are present in stretched state only without oscillation therefore Brahamand is in unique state named as hybrid dark matter –energy state. During the formation of Brahamand when the the centres of tetrads just touched surfaces of shells and the radius sphere of tranquillity reached the minimum value the Brahamand exploded into two parts with big bang. The two parts are named as anantrons because each is precursor of infinite mass and energy. Ananta means infinite in Sanskrit. Each anantron is composed of triads of shells. Anantron composed $A^{\wedge}(\pm)A(+)$ is called $(1+)$ and anantron composed of $A^{\wedge}(\pm)A(-)$ is called $(1-)$. The anantrons are composed of stetched triads therefore during big bang anantrons flew apart with very high velocity even greater than velocity of light which was possible because anantrons were bereft of mass like properties. The energy of explosion imparted oscillatory motion to the stretched shells at the stage when the two anantrons were separated by very huge distance and thus anantron$(2+)$ and $(2-)$ with properties of mass were born. The anantron$(2+)$ is source of very huge number of neutrons while anantron$(2-)$ is source of anti neutrons. The two types of anantrons gave birth to matter and anti matter $A^{\wedge}(\pm)A(+)$ changed to neutrons which is further source of protons and electrons while

A^(±)A(−) was source of anti neurons which produced antiprotons and positrons. The matter and antimatter were produced not at the same spot were evolved after the separation of two types of anantrons by very very huge distance, therefore, coming together of matter and antimatter universes are very slim. The big bang was not clean process therefore very huge number mass and charge shells lumps in stretched state only without any oscillations were littered around. The lumps are source of dark matter and dark energy.

The black holes are akashons composed of tetrad shells. The minimum radius of sphereof tranquillity of an akashon is equal to radius of the shells of which the akashon is composed of hence very huge mass is associated in black hole akashon because it is composed of shells having radius nearly equal to zero. The con ventional akashon is composed of hybrid charge – mass shells having radius much greater than tetrad shells hence mass of conventional akashon cannot be greater than N-akashon. In the beginning there was an Akashon with very very very large radius of sphere of tranquillity. As already seen mass, energy and time period measured by using celestial time are inversely proportional to radius of sphere of tranquillity. The radius of sphere of tranquillity is very very large, therefore in the initial stages all the mass, energy and time were nearly zero. Nothing is born except the very very big Akashon or cosmic egg. There was very very little motion in the region of perturbation. When two cosmic eggs composed of Akasha and anti-Akasha shells having mutual attraction happen to come close the creation of matter and energy along with the field and time started. As the cosmic eggs came closer and closer the oscillatory tranverse motion of the shells present in the region of perturbation increased giving birth to mattenergy and time. The radii of the shells present in the region of perturbation in the initial stages was very very small. The energy generated due to attractive force was stored up in the region of perturbation and hence radii of the spheres of tranquillity of both the cosmic eggs became smaller and smaller till the radii of spheres of tranquility became equal to the radii of shells. At this stage the two cosmic eggs fused together to form primevil Akashon with very small radius of sphere of tranquillity and hence very very huge mass. All the mass was concentrated in nearly point volume. Simultaneously with the merger of two cosmic eggs the attractive force of the eggs changed into repulsive force which led to big bang or scttering around of energy and matter in the form of Akashons or Prakashons. It is not necessary that all of it happened at one point. It might have happened at many points and the process of creating might be taking place at present at other points also. The expansion process may again change into reverse contraction process.

How much mass is to be associated with the Akashon depends on the radius of sphere of tranquillity. The gestation of the Akashon involves the inward displacement of shells towards the evolving mass center. As already seen the displacement is inversely proportional to the distance of the shells from the mass center. The limiting displacement stage is reached when the displacement is equal to the radius of the shells because more displacement will shift the center of the shells outside it but it is not allowed. If very huge numbers of shells glue together to form Akashon, it is evident that the innermost shells situated on the periphery of the sphere of tranquillity will acquire the limiting value of displacement, and when it happens the nature of Akashon shell changes into that of Prakashon (Photon) shells. The shells following the innermost shells are pushed inward to occupy the slots vacated by the metamorphosed shells and hence meet the same fate as the previous shells. The metamorphosis process is completed in very very short time and hence it is catastrophic therefore, huge amount of energy is accumulated in nearly point space which leads to big bang.

DARK MATTER AND DARK ENERGY:

When the innermost shells after attaining the limiting displacement undergo conversion as already discussed the other following shells also meet the same fate. Extra energy is liberated when the shells moves. As already seen the cosmic force acting on each shell is equal to $k_F/4\pi r^2$ if it is supposed that when the limiting displacement is reached the radius of the sphere of tranquillity is equal to r, which is also equal to the radius of each shell, from simple geometry it is simple to see that the innermost shell waiting to occupy the vacated slot drops through distance 2r. Therefore the energy liberated by each shell is equal to $k_1/2\pi r$ because maximum force achieved with maximum displacement remains constant. The number of shells in each Akashon is equal to k_2/r and the total energy is equal to $k_1 k_2/2\pi r^2$. The mass or the energy associated with the photon or Akashon is equal to k'/r, it is evident that the energy liberated during inwards motion of the shells is much more than the energy congealed in the photon or Akashon because r is very very small. The extra energy is strewn around as dark matter which is composed of unbaked Akashon in which the shells are stretched but lack the independent oscillations. The gravitational effect is there due to stretching as will be shown later on but on account of absence of independent oscillations the usual interaction effects of complete Akashon are missing.

It can also be said that dark matter is composed of Akasha shells arranged around the sphere of tranquillity like full fledged Akashon but shells are only stretched without any oscillations. In the same manner, dark energy is composed of stretched anti-Akasha shells without oscillating shells. Therefore, dark matter and dark energy are uncooked Akashon and

Evolution of Physical Laws

Prakashon. Due to stretched shells dark matter and dark energy are associated with force property. The amount of dark energy is more than that of dark matter because most of the Akasha shells are used to form Akashons which is representative of conventional matter but the anti-Akasha shells are more abundant. The missing anti-matter leads to the abundance of anti-Akasha shells.

The dark matter is composed of doublet shells of positive and negative Akasha in stretched state but without any oscillations. Dark energy is composed of doublet shells of positive and negative anti-Akasha in stretched state without any oscillation. When dark matter and dark energy combine the exchange of shells lead to the formation of conventional matter.

THE BLACK HOLE:

The black holes are massive Akashons where the innermost shells have not attained the limiting displacement. The apparent big size of the black holes is due to the accumulation of mass around the black holes which is to be sucked in by it because assimilation is a very slow process. It is so because before eating up, the complex matter has to be broken down to simpler Akashons. With the intake of more and more mass by the black holes, the innermost shells are pushed inwards and at certain stage limiting value of displacement is reached which leads to big bang. The cosmic show of birth and death is eternal.

THE TRUE AKASHON:

The so called elementary particle is said to be true. Akashon if the mass equation is applicable to it. If the electron is taken as true Akashon, the radius of the sphere of tranquillity is given by

$r = h / 4\pi m.c$

putting the values of the constants

$r = 1.93 \times 10^{-13} m$

Similarly if the proton is considered as the true Akashon, the value of the radius of the sphere of tranquillity is 1.053×10^{-16}. The value of the radius of the sphere of tranquillity of the electron is much bigger than that of the proton. In case of the ground state of the hydrogen atom, the proton is lying snug within the sphere of tranquillity of the electron or the sphere of tranquillity of proton lies inside the sphere of tranquillity of the electron. The composition of hydrogen atom will be discussed later on also where it shall be proved that the ground state of the hydrogen atom is the same as pictured above.

The value of energy density in the region of perturbation of the Prakashon can be determined in the same manner as in case of Akashon.

$\rho_E = (hc/16\pi^2).(1/r^4) = (1.258/r^4) \times 10^{-27} J.m^{-3}$

The energy density is inversely proportional to the fourth power of the distance but unlike the mass density the fall in the energy density with increase of distance is not so steep because the value of the constant is

c^2 times the value of constants related to the mass density, consequently the Prakashon is more spread out than the Akashon.

$$\rho_E = \rho_m.c^2$$

The number density of the shells present in the region of perturbation of the Akashon (mass packet) and Prakashon (energy packet) can also be determined. The limiting number of shells present in the Akashon and the Prakashon are given by the principle of inverse variation.

$4\pi r.N_m = k_{N(m)}$

$4\pi r.N_E = k_{N(E)}$

The number densities of the displaced shells present in the Akashon (mass packet) and the Prakashon (energy packet) are given below

$\rho_{N(m)} = k_{N(m)} / 16\pi^2 r^4$

$\rho_{N(E)} = k_{N(E)} / 16\pi^2 r^4$

As already pointed out, the Prakashon (photon) is more spread out than the Akashon (mass packet), therefore, the number density of the Prakashon does not decrease so rapidly as the number density of the Akashon. The above density is possible only if $k_{N(E)}$ is more than $k_{N(m)}$

$4\pi r.N_E = k_{N(E)}$

$4\pi r.N_m = k_{N(m)}$

$k_{N(E)}$ is more than $k_{N(m)}$, therefore, $N_E > N_m$. Thus the Prakashon contain more number of displaced shells than the Akashon having the same radii of

Evolution of Physical Laws

the spheres of tranquility. The average energy and the average energy mass per shell packed in the Akashon and the Prakashon can also be calculated.

$dE = - h.c / 4\pi r^2$

$dN_E = - (k_{N(E)} / 4\pi r^2).dr$

dE and dN_E are the energy and the displaced shells present in the volume $4\pi r^2.dr$.

average energy per shell = $hc / k_{N(E)}$

similarly, average mass per shell = $dE / dN_E = h / c. k_{N(m)}$

As already seen the mass and energy are inter convertible, therefore, in case of the Akashon instead of calculating the average mass stored per nodule, the energy equivalent of mass can be used. $dE = dm.c^2$. The average mass stored per shell is

$dm / dN_m = k / c. k_{N(m)}$

When both sides are multiplied by c^2, the average mass changes to energy, average energy per shells becomes

$dm.c^2 / dN_m = dE_{(m)} / dN_{(m)} = h.c / k_{N(m)}$

dE(m) denotes the energy equivalent of the mass.

As seen earlier the average energy stored per shell in the Prakashon is

$dE/ dN_{(E)} = h.c / k_{N(E)}$

It is also known that $k_{N(m)} < k_{N(E)}$. Therefore, the average energy per shell in the Akashon (mass packet) is more than the average energy per shell in the Prakashon (energy packet), if the Prakashon having same radius, the energy of the Akashon is now is to be distributed amongst displaced shells of the

Prakashon. In the other words it can be said that the Akashon (mass packet) is the condensed form of the Prakashon (photon) or Prakashon is the bloated form of the Akashon. It seems probable that at one stage, the Universe was so much saturated with the Prakashons that the Prakashons precipitated as Akashons in the same manner as the salt is precipitated from the saturated solution. The nature of dark matter will be discussed during discussion on gravity. It has been seen that theoretically sphere of influence of photon extends up to infinity. The presence of pair of entangled photon is proof that actually it is so otherwise the photon cannot influence each other when the photons are located at large distance from each other.

SPECIAL THEORY OF RELATIVITY:

The backbone of special theory of relativity is not invariant value of c but the relation $r = r_0\sqrt{\{1 - (v^2/c^2)\}}$ where r_0 is the radius of an Akashon at rest and r is the radius when the same Akashon is moving with velocity v. The radius relation was based on equivalence of mass and energy which is logical because the progenitor of mass and energy is the ether. The common source of manifestation of mass and energy leads to the logical conclusion that mass and energy are equivalent. The above relation has been deduced without using bizarre concept of Lorentz's contraction. It has also been seen that the invariance of c is mathematical necessity, being constant of proportionality having dimensions of velocity. The concept of frame is no longer needed. Special theory of relativity results are based on subjective observations but science is based on objective results independent of subjective interpretation. How the space associated with the moving bogeys of train can be equated to space with train at rest! It is only an optical illusion due to image on retina moving in opposite direction that even the stationary train appears to be moving which leads to illogical conclusions based on subjective observation that the space of two trains is equivalent. Actually that train is moving on which force has been impinged irrespective of visual observation and independent of observer. The science is not based on illusion and that is why the results of special theory of relativity based on illusionary concepts seem to be illusions. All the results of special theory of relativity based on bizarre concept of frames except mass energy equivalence are redundant. All the results of special theory of relativity are packed up in $r=r_0\sqrt{\{1-(v^2/c^2)\}}$, which was derived in simple manner. The derivation was based on logical concept of equivalence of mass energy.

Evolution of Physical Laws

The cornerstone of the special theory of relativity is the invariance of the speed of light, without considering why it is so but in the present text the treatment is such that invariance of c is only corollary and not the seed. c is benchmark velocity which happens to be equal to the velocity of light and it was introduced out of mathematical necessity. It has already been proved that the velocity of the Akashon with zero mass is the maximum and invariant because it is the special value of the terminal velocity of an Akashon as got from the general terminal velocity relation.

In the treatment so far presented, the two most important results of the special theory of relativity, the effect of motion on the mass

$$m = m_0 / \{\sqrt{(1-(v^2/c^2))}\}$$

and $E = mc^2$ has been derived without using Lorentz's transformations. The method is simple and explicit and makes use of the principle of inverse variation and the model of the Akashon in combination with the concept of force and energy. The extra mass created as result of motion is temporary mass stored in the temporary region of perturbation but it has got the same properties as that of permanent mass.

Relativistic mass $m = h/(4\pi rc)$

Momentum $p = mv = hv/(4\pi rc)$

$r = r_0 \{\sqrt{(1-(v^2/c^2))}\}$

$m^2 = h^2/16\pi^2 r^2 c^2 = h^2/[16\pi^2 r_0^2 \{(1-(v^2/c^2))\}.c^2]$

$(m^2c^2) \cdot \{(1 - (v^2/c^2)\} = h^2/16\pi^2 r_0^2$

$m^2c^2 - m^2v^2 = h^2/(16\pi^2 r_0^2)$

Multiply both sides by c^2,

$m^2c^4 - m^2v^2c^2 = h^2c^2/\{(16\pi^2 r_0^2)$

$p = mv$

$(mc^2)^2 - (pc)^2 = h^2c^2/\{(16\pi^2 r_0^2)$

m_0 is the invariant mass

$h^2/(16\pi^2 r_0^2) = m_0^2 c^2$

$(mc^2)^2 - (pc)^2 = m_0^2 c^4$

$E = mc^2, E_0 = m_0 c^2$

$E^2 - (pc)^2 = E_0^2$

That the velocity of light is invariant is not fundamental concept but the more general concept is the principle of inverse variation because with the models of Akashon (mass packet) and the Prakashon (energy packet) based on this principle, the results of the special and general theory of relativity can be derived without using any complicated mathematical tools. There is no need to introduce the concepts, which do not appeal to the common sense and cover the theory of relativity in enigmatic veil of mathematical rigamarole beyond the comprehension of the ordinary man. The importance of higher mathematics in understanding the secrets of nature is not belittled here but if possible; simple way can also be adopted. Sometime the mother nature can open doors to reveal its secrets by merely gentle knocking.

RELATIVE MEASUREMENT OF LENGTH AND TIME:

In thought experiment length is specified by fixing two imagenary points in space and an imaginary clock is used to measure time, but in real experiments imaginary specification of length and time is not suitable. It will be discussed now that how practical natural units of length and time can be specified. Akashon can be used as natural meter rod and natural clock for the measurement of length and time. The basic equation for this purpose is $4\pi r v$ or $4\pi r/t = c$, t is the time required to complete one oscillation by the shell located on the surface of sphere of tranquillity of radius r. r and t are used as natural standard units of length and time

$r/t = c/4\pi$

The equation shows that while fixing the units of length and time the fixing of time and length unit is related by above equation, therefore, **arbitrary fixing of length and time units is not natural.** According to natural units when time t = 1, length unit will be $c/4\pi$.

Measurement of length:

The length of material rod cannot be used to specify standard unit of length because as already discussed, ends of the meter rod are not sharp but fuzzy hence, fixing of the length of rod is not accurate. The diameter or radius of sphere of tranquillity can be specified as natural accurate unit of length because surface of sphere of tranquillity can be located exactly because it is not fuzzy.

Suppose radius of sphere of tranquillity at rest is r_0 and it changes to r when the Akashon is subjected to force to acquire velocity v. As already seen,

$r = r_0 \{\sqrt{(1-(v^2/c^2))}\}$

Suppose, n spans of standard unit of length r are marked by using moving Akashon in moving frame,

$L = n.r$

Next suppose, same Akashon at rest is used to span n units of length,

$L_0 = n.r_0$

$L/L_0 = r/r_0 = \sqrt{(1-(v^2/c^2))}$

$L < L_0$

So there is length contraction in moving frame.

No experiment is feasible to measure length contraction. If a rod is selected for length measurement experiment as already mentioned it is not possible to specify the exact length of the rod. Suppose, in thought experiment a rod is selected which is composed of a number of spheres of tranquillity lined together touching each other but practically if such a rod is made to move, the core of the Akashons i.e. sphere of tranquillity will only contract but there will be no change in the length of the rod as a whole.

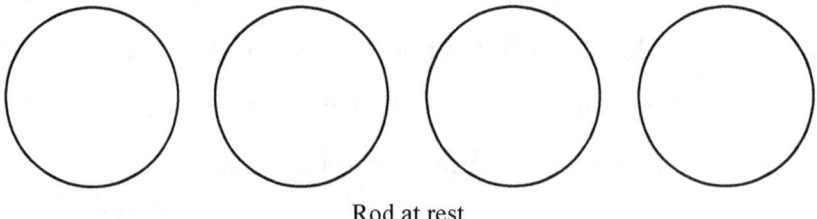

Rod at rest

Fig. 10

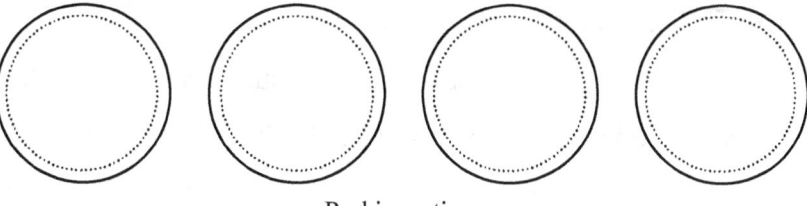
Rod in motion

Fig. 11

The fuzzy ends of a real material rod will not become closer while in motion because it is only the inner radii of spheres of tranquillity of Akashons of which matter is composed that contract. The contraction of inner core of the Akashon can not be measured. All the paradoxes associated with length contraction are redundant.

One unit of length of stationary frame = $\sqrt{(1-(v^2/c^2))}$ unit of length in moving frame in terms of stationary units. The unavoidable change (contraction) in the radius of sphere of tranquillity of stationary Akashon when shifting to moving frame as standard unit of measurement of length lead to the length measurement discrepancies, **in reality the empty space does not contract.** Even the material rod does not contract while in motion, it is only the inner core of the Akashon which contracts without changing the length of rod. The length of rod will not become zero when $v = c$, though radius of sphere of tranquillity may be equal to zero.

No experiment can be set up to actually measure the length contraction.

MEASUREMENT OF TIME:

The Akashon can be used as natural time measurement clock. The time needed to complete one oscillation by a shell situated on the surface of sphere of tranquillity can be specified as cosmic unit of time or a number of oscillations completed in time interval taken as one unit of time can also be celestial unit of time. For specifying unit time, the following equation based on principle of inverse variation is used:

$4\pi r \nu = c$

Suppose, the time interval during which ν number of oscillations are completed is taken as one unit.

In stationary frame of reference the above equation is $4\pi r_0 \nu_0 = c$, because r is less than r_0 it means ν is greater than ν_0, which shows that the greater number of oscillations will have to be fixed for unit time interval in moving frame as compared to stationary Akashon clock or it can be said that moving Akashon clock is retarded.

If t is the time required to complete one oscillation in moving Akashon clock, ν will be equal to $1/t$.
Similarly, for stationary Akashon clock $\nu_0 = 1/t_0$.

$4\pi r/t = c,\qquad t \propto r$

$4\pi r_0/t_0 = c,\qquad t_0 \propto r_0$

$r < r_0$, therefore $t < t_0$ which shows that moving Akashon clock unit of time is less or it can be said that it is retarded.

Time flow can be compared to flow of water in pipe; t and t_0 are rates of flow of time as measured by the stationary Akashon and moving Akashon respectively.

$t = 4\pi r/c$

$t_0 = 4\pi r_0/c$

The time T recorded by moving Akashon clock for n ticks = n.t

The time recorded for n ticks by stationary clock T_0 = n.t_0

$T/T_0 = t/t_0 = r/r_0 = \sqrt{\{1 - (v^2/c^2)\}}$

$T < T_0$

This shows that moving Akashon clock is retarded.

Rate of flow of time = Time required to complete one oscillation.

Rate of flow of time α radius of sphere of tranquillity of Akashon clock.

t can also be taken as one unit of time i.e. time required for completing one oscillation by shell. As $r_0 > r$, it means $t < t_0$ or it shows moving Akashon clock is retarded.

$t/t_0 = r/r_0 = \sqrt{(1 - (v^2/c^2))}$

The above equation shows that one unit of time shown by the stationary clock is equal to $\sqrt{(1-(v^2/c^2))}$ unit of time as measured by movng Akashon clock.

How much water flows through pipe during given interval α rate of flow of water.

How much time is needed to fill the bucket or to complete a process α 1/(rate of flow of water or time or unit time).

Time required to complete a process as measured by the given Akasohon clock is 1/(rate of flow of time).

Suppose, T' and T_0' are time measured by moving Akashon clock and stationary Akashon clock r and r_0 are the radius of the Akashon clocks respectively. For the same process to be completed.

T' α 1/t α 1/r; T_0' α 1/r_0 α 1/t_0

T'/T_0' = r_0/r = $1/\sqrt{\{1-(v^2/c^2)\}}$

T' > T_0', this shows moving Akashon is retarded

The atomic clocks are equivalent to Akashon clocks. It is actually seen that when moving atomic clock is used in GPS the moving atomic clock in the satellite looses timedue to slow ticking as compared to the stationary atomic clock on the ground. The satellite clock needs setting. The reverse is the case when the moving Akashon clock and the stationary Akashon clock are used to measure time for the completion of the same process. If T and T_0 are the times measured by the moving and stationary clocks respectively,

$T \alpha\ 1/t\ \alpha\ 1/r$

$T_0 \alpha\ 1/t_0 \alpha\ 1/r_0$

$T/T_0 = r_0/r = 1/[\sqrt{(1-(v^2/c^2))}]$

$T = T_0/[\sqrt{(1-(v^2/c^2))}]$

$T > T_0$

For infinitesimle time interval, $dt = dt_0/[\sqrt{(1-(v^2/c^2))}]$

The above result is observed while measuring the lifetime of moving muon and stationary muon because the time required is for the same process to be completed. The life of moving muoon is more as compared to the stationary muon because moun behaves like a moving Akashon clock. The rate of disintegration of moving Akashon is slowed down as compared to the stationary muon. Similarly it can be said that rate of chemical reaction wil be less if the raction is taking place in moving frame as compared to the stationary frame because reactants' reaction rate is governed by the Akashon clocks associated with the reactants. In human body so many chemical reations are taking place, so it can be inferred that rate of aging is retarded in the moving frame. **The discrepancies observed in the time measurement are due to the unavoidable changes in the standard units of time as specified by the moving and stationary frame.**

Evolution of Physical Laws

Further views on length and time measurement (Space-Time Continuum):

The measurement of length (space) and time are not independent if natural standard units of length which are incorporated in the Akashon are used as standards. The methods to measure time and length are inter-locked in the equation $4\pi r\nu = c$ or $4\pi r(1/t) = c$, space α time. The ticking of time started only after the birth of Akashon (matter) and Prakashon (energy). It cannot be said that there was no time prior to the gross manifestation or big bang.

The relative time and space are born with the birth of Aklashon (big-bang). The absolute time and space were present even before the big-bang but the natural metre and clock, the Akashon was not there. The measurement of space and time is not possible in the absence of Akashon, the natural meter rod and natural clock. The big-bang gave birth to Akashon and not the absolute space and time. With the birth of Akashon, measurement of space and time started.

Measurement of velocity of light:

The following equation as already mentioned is associated with the Akashon and Prakashon $4\pi r\nu = c$ or $4\pi r.(1/t) = c$. The simple equation relates space and time. The length and time measurement units are $4\pi r$ and $1/t$ (i.e. ν) respectively. Suppose radiation starts from a shell. It is self evident from the above equation that wavelength and frequency of radiation will be $4\pi r$ and $1/t$ (i.e. ν) respectively, therefore, product $4\pi r\nu = 4\pi r.(1/t) =$ velocity of radiation = constant = c. The above result proves that velocity of light is independent of frame velocity. Although $4\pi r$ and $1/t$ change with the state of motion of the Akashon or frame, the product of these two quantities is constant because, the magnitudes of these two quantities are related according to the fundamental principle of inverse variation. Now it is not difficult to understand the reason for the invariance of velocity of light. The above property of light gave negative results in Michelson-Morley experiment but the experiment does not deny the presence of ether because Michelson-Morley experiment was faulty. It was based on prior assumption that light velocity changes with the frame velocity which was based on erroneous concept that ether is stationary. The external terrestrial units of length and time do not spontaneously synchronize with the internal celestial units associated with the Akashon and hence are unsuitable.

All the paradoxes associated with the relative measurement of time and length are absurd because as already discussed that there is no contraction of the length of the rod, it is only the temporary inwards growth of the region of perturbation which leads to the apparent discrepancies because the observers are unaware of this change. The actual dimensions of moving solid objects do not change. The change in the half-life of unstable particles while in motion is due to creation of extra region of perturbation which decreases the radius of sphere of tranquillity of Akashon clocks of which the particles are composed and hence rate of disintegration is affected.

Finally it is concluded that it is the principle of invers variation $4\pi r v = c$, on which special theory of relativity is based. The invariance speed of light can be side tracked. No frame concept or subjective observations are needed to get the results of special theory of relativity. If one frame is actually moving towards right after it has been subjected to force, it does ont mean that the other frame unrelated to the first frame is moving towards left because that is only an optical subjective illusion as in case of moving train and the platform moving opposite direction. If the principle of inverse variation is followed there is no need to discard **common sense** to understand theory of relativity.

Evolution of Physical Laws

GENERAL EXPRESSION FOR THE MEASUREMENT OF LENGTH AND TIME IN THE TWO FRAMES MOVING WITH RELATIVE VELOCITY:

Suppose frame S' is moving with velocity v relative to the frame S along the X-axis. Suppose co-ordinates of any point P fixed in space are to be determined along X-axis at any instant in the two frames.

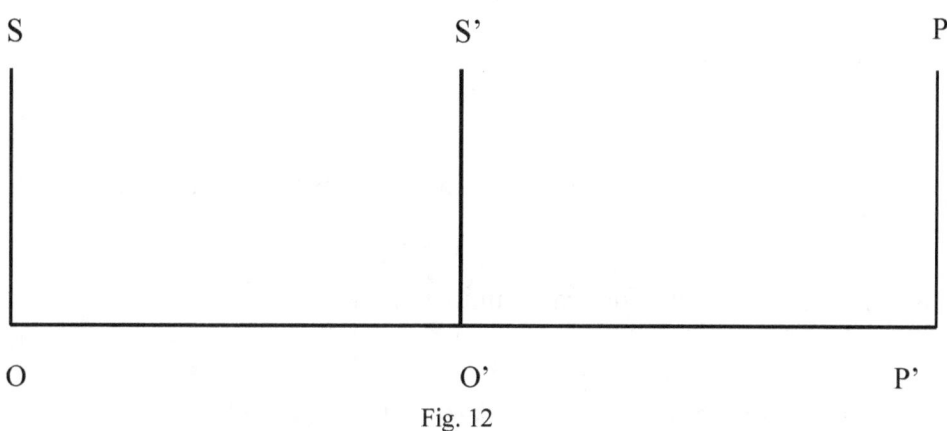

Fig. 12

The Akashon clocks fixed at each point in space are used to measure time. O and O' are the origin in the stationary and moving frames respectively. When O and O' overlaps, zero time is recorded on clocks. Let x be distance of the point P along X-axis from the origin of the stationary frame as measured by the standard natural length unit as specified by the frame at rest. Suppose during time t as measured by the clock at rest, origin O' of the moving frame S' reaches point P' marked on the X-axis of the frame at rest.

O'P' = (x – vt) as measured in standard rest units,

O'P' is also = x' as measured in specified moving frame units,

As already discussed, one specified unit of length measured in the moving frame unit = $1 - v^2/c^2$ units when converted to standard specified unit of frame at rest.

Therefore,

x' units of moving frame = $x' \sqrt{(1 - (v^2/c^2))}$ units of frame at rest.

$$x' \sqrt{(1 - (v^2/c^2))} = (x - vt)$$

$$x' = (x - vt) / [\sqrt{(1 - (v^2/c^2))}]$$

Suppose next moving Akashon clock placed at O' is used as time measuring device.

OP when measured in moving frame units will be,

OP = x' + vt'

OP = x as measured in rest frame standard units

Rest frame units = $\sqrt{(1 - (v^2/c^2))}$ moving frame units.

Therefore,

$$x \cdot [\sqrt{(1 - (v^2/c^2))}] = x' + vt'$$

$$x = x' + vt' / [\sqrt{(1 - (v^2/c^2))}]$$

By putting the value x from second relation in the first relation and again putting the value of x' from the first relation in the second relation, the apparent time relations are got.

$$t = t' + (xv/c^2) / \sqrt{\{1-(v^2/c^2)\}}$$

$$t' = t - (xv/c^2) / \sqrt{\{1-(v^2/c^2)\}}$$

The special theory of relativity is based on the exact measurement of the radius of the sphere of tranquillity. It will be seen later on that there is inherent uncertainty in the measurement of radius which Einstein refused to admit, however the uncertainty or error is very very less.

The above basic result of the special theory of relativity has been derived without using the concept of the invariance of speed of light because the principle of inverse variation is more fundamental concept which leads to invariance of light velocity. When two frames are moving only that frame on which force has been impressed has got real motion. The relative motion of the other frame is only illusionary and depends upon the eye of the observer in the first frame. Therefore, the motion of the second frame on which force does not act is only illusionary and subjective but study of science is not subjective.

The paradoxical results of the special theory of relativity connected with the measurement of length and time are actually the effects of the shorter radius of the sphere of tranquillity of the Akashon in motion as compared to the Akashon at rest. The special theory of relativity is based on subjective observations such as simultaneous observation of phenomenon by two observers but in science subjective experience does not carry much weight.

The way the relativity has been discussed in the present text is without any subjective conclusions.

The Michelson-Morley experiment which is connected with the special theory of relativity will be examined later on from another angle while discussing the dual nature of electromagnetic radiation.

The model of the Akashon is based on the principle of inverse variation. The principle of inverse variation can also be named as the principle of unification as shall be seen later on when with the help of model of the Akashon, many unrelated field of study are unified and the application of the model concept of the Akashon result in deriving the proofs of the so called basic relations connected with unseemingly unrelated natural phenomena. It shall be seen that why some of the basic measurements fit together in specific pattern to form an equation which is actually mathematical tool for knowing the value of another basic property of nature without measuring it such as the Newton's Law of gravitational force.

CONCEPT OF FIELD:

So far the paradigm field concept is that each field is associated with field carrying particles virtual or real Higgs has gone so far that it is not only special particle's existence that depends on the corresponding associated field and the prime cause of all the matter in the universe is omnipresent elusive Higgs field. The manifested mass is floating through the Higgs field. No Higgs field, no mass. There are serious drawbacks in the above hypothesis. According to me as already discussed, the matter is an illusion which is the effect of oscillations of shells present as packet in Akashon. No oscillations, no mass. The oscillations need force or it can be said that cause of mass manifestation is the cosmic force. The strong cosmic force is omnipresent like the Higgs field but with varying intensity. The cosmic force is concentrated where Akashon is present. The appearance of net force during particle interaction is due to the unbalancing of cosmic force. The unbalanced cosmic force is associated as field during interaction. The cause of all the fields gravitational, electromagnetic, weak and strong force is unbalanced cosmic force. The strain in the Akasha (ether) is the cause of field. The concept of field carrier cannot lead to comprehensive derivation of Coulomb's law and Newton's relation for gravitational force. The new concept of field can be used to deduce the above mentioned force relations.The field is produced due to elastic stretching or compressing of springs of ether or anti-ether while the casue of mass formation or energy formation is due to oscillations of shells of ether or anti-ether. With mass packet or energy packet field will always be associated but it is not necessary with all the fields mass packet or energy packet is associated. Nothing is carrier of field like photon, gluon etc. By using concept of virtual particles as carriers of field comprehensive derivation of Coulomb's Law or Newton's Gravitation Law involving charge-mass distance cannot be calculated.

THE MANIFESTATION OF GRAVITATIONAL FIELD:

The novel field theory presented here does not need the bizarre and unrealistic concept of virtual particles or space bending. The gravitational and electrostatic or electromagnetic fields can be explained without introducing the hypothetical concept of virtual graviton or virtual photons. The Akashon (etheron) is composed of displaced shells which are arranged in the region of perturbation around the sphere of tranquillity. The shells line up to form springs which start from the sphere of tranquillity and go upto infinity. In case of Akashon, the springs are compressed inwards. The force is acting on each shell of the shell spring due to which centre of shell is shifted away from the geometric centre in a direction away from the direction of force. The restoring force acts on the displaced shell center in direction opposite to the squeezing force acting on the spring shells. The shift is away from the geometric centre of the etheron. As the distance of the shell center in the spring becomes more from the geometric centre of the etheron, the shifting of the shell centre becomes less and less because as the distance increases, force becomes less and less. In case of an absolutely isolated Akashon no resultant cosmic force is there because springs are being comperssed inward equally around the sphere of tranquillity leaving no net cosmic force. But when two Akashons interact or are brought near to each other, the cosmic force is unbalanced; this unbalanced cosmic force is manifested as the gravitational field. The mutual interaction or **spooky action** at a distance involving Akashons is possible at all the distances because the region of perturbation of the Akashon is boundless. It has been seen that the Akasha distributed over the shellsurfaces is not uniform. The more dense regions will be shown with positive sign and rare region will be

Evolution of Physical Laws

shown with the negative sign during interaction. While showing interaction of two Akashons only limited number of springs attached to the sphere of tranquillity will be depicted.

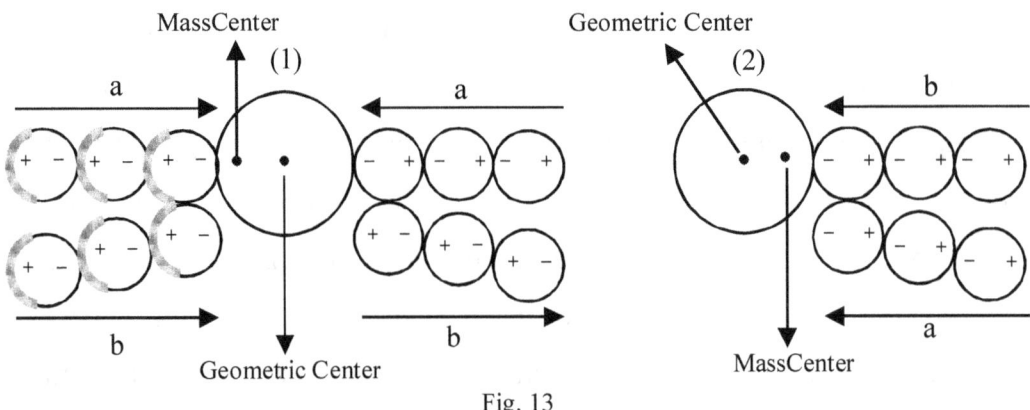

Fig. 13

1. Let us next discuss the mutual interaction of the Akashons. In the above diagram big circles (spheres) are the spheres of the tranquility of Akashon (1) and Akashon (2). Small circles are shells of theether. The truncated springs belonging to Akashon (1) and Akashon (2) respectively are shown by 'a' and 'b'. In a completely isolated etheron, no resultant cosmic force is acting on the etheron. When two etherons or Akashons interact, the cosmic force is unbalanced which manifests itself as gravitational force. The force was already there but it was unbalanced to appear as attractive gravitational force on mutual interaction. As seen in the diagram, in the left part of the region of perturbation Akashon-1 dense parts of shells 'a' and 'b' overlp, which increases the disproportionate distribution of Akasha over the shell surface, hence compressive force of the shell springs acting from left to right is increased as compared to force when Akashon-1 is isolated.

Similarly in the right portion of region of perturbation of Akashon-1 the denser part of the shells belonging to 1overlap with the rarer part of the shells belonging to Akashon-2, hence decreasing the disproportionate distribution of Akasha which results in less force of compression acting on Akashon-1 from right to left. Overall result of interaction is that Akashon-1 experience net pushing force from left to right or it can be said that Akashon-2 is attracting Akashon-1. As a result of net force mass center of Akashon-1 is shifted to left. By the same arguments it can be shown that Akashon-2 is attracted by Akashon-1 with shift in mass center towards right. The net result of interaction is that Akashon (1) and (2) appear to be attracting each other or unbalanced cosmic force here appears as gravitational force.

The centre of each shell (mass centre) towards left of the Akashon (1) shifts more towards left due to reinforcing of the force and reverse is the case in the region right to the Akashon (1). As there is shifting in the mass centres associated with the shells, the mass centre of the sphere of tranquility or of the Akashon as a whole will also shift towards left of the geometric centre. During this interaction there isno change in the geometric center of the shells, therefore there is no shift in the geometric centre of the Akashon.

2. Apart from change in the displacement of shell, slight change in the number of shells associated with the interacting Akashons also contributes to the manifestation of the attractive force. The sphere of tranquillity of the Akashon (1) lies in the region of the perturbation of

Akashon (2) and vise-versa. No shells can be present in the sphere of tranquillity, therefore the shells present inside the (1) and (2) spheres of tranquility which are associated with the other Akashon are ejected out and the expelled shells are accommodated on the surface of the sphere of tranquillity of Akashon (1) near point L and in the case of (2) near point M, because these regions are comparatively less crowded as compared to region P.

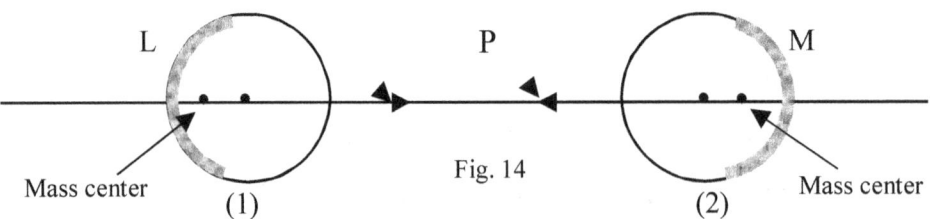

Fig. 14

After ejection from the sphere of tranquillity (1) and (2), the shells are again attracted towards the center of the Akashon. The net result of reallocation of the shells results in the shifting away of the mass centres of both the Akashons.

The above discussion is simplified explanation of the mutual interaction of the Akashons and there might be other mind-boggling explanations but one thing is clear the **cosmic force is already there to be manifested as secondary force called gravitational force** on mutual interaction. The manifestation of the gravitational force is not instantaneous process, because the readjustment of the shells need sometime.

DERIVATION OF NEWTON'S GRAVITATIONAL FORCE RELATION:

Let us next discuss the magnitude of the gravitational force. The mutual interaction of the Akashons leads to force acting on the Akashons, therefore, the mass centers of the Akashons are shifted away from the geometric center. Suppose l is the distance of separation of the mass center and the geometric center of an Akashon and r is the radius of the sphere of tranquility. The force experienced by an Akashon is given by the following usual Hooke's force relation

Force = $F = F_0 . l / r$

The force factor F_0 here is the result of mutual interaction therefore, the value of F_0 depend on the extent of mutual interaction or distance between the interacting Akashons. It will be seen shortly that the force factors remains same for all the Akashons having same distance of interaction.

Suppose Akashons having masses M_0 and m_0 are placed at the same distance from another Akashon having some mass. M is the mass of the single Akashon which is equivalent to the two Akashons. r, r_0 and r'_0 are the radii of the spheres of tranquillity of the Akashons having masses M, M_0, m_0 respectively. The force acting on the equivalent Akashon when it interacts with given Akashon is equal to the sum of forces which act on the individual Akashon during interaction with the same Akashon as chosen for the equivalent Akashon.

$M = M_0 + m$

$$h / 4\pi rc = h / 4\pi r_0 c + h / 4\pi r_0' c$$
$$1/r = 1/r_0 + 1/r_0'$$

$$F_1 = F_{0(1)} \cdot (l_1 / r_0); \quad F_2 = F_{0(2)} \cdot (l_2 / r_0'); \quad F_3 = F_{0(3)} \cdot (l_3 / r)$$

l_1, l_2, and l_3 are displacements of mass centres of M_0, m_0 and M respectively. The force factors $F_{0(1)}$, $F_{0(2)}$, and $F_{0(3)}$ are equal because the interacting distance between the Akashons is same. The total force acting on single Akashon is equal to the sum of two forces acting separately on two parts of the Akashon.

$$F_3 = F_1 + F_2$$
$$F_{03} \cdot (l_3 / r) = F_{01} \cdot (l_1 / r_0) + F_{02} \cdot (l_2 / r_0')$$

The force factors are equal
$$F_{0(1)} = F_{0(2)} = F_{0(3)}$$
$$l_3/r = (l_1/r_0) + (l_2/r_0')$$

The mass of an Akashon is inversely proportional to its radius of sphere of tranquillitym, therefore
$$1/r = k.M' \qquad 1/r_0 = km_0 \qquad 1/r_0' = km$$

put the values in the above relation
$$l_3.M = l_1.M_0 + l_2.m_0$$

If the displacement of the mass centre of an Akashon having gravitational interaction is supposed to be indirectly proportional to its own mass which means, $l_3.M = l_1.M_0 = l_2.m_0 = k_1$ and hence $k_1 = k_1 + k_1$ but it is not true.

Let us next suppose that displacement is directly proportional to the respective masses, therefore,

$l_3 = k'M$

$l_1 = k'M_0$

$l_2 = k'm_0$

k' = Constant

After putting the value in original relation,
$M^2 = M_0^2 + m^2$

But the above relation is not true. Therefore, the displacement of the mass center is not directly proportional to its own mass. So finally it is concluded that the displacement of the mass centre of an Akashon during gravitational interaction is independent of its own mass but depends on other factors.

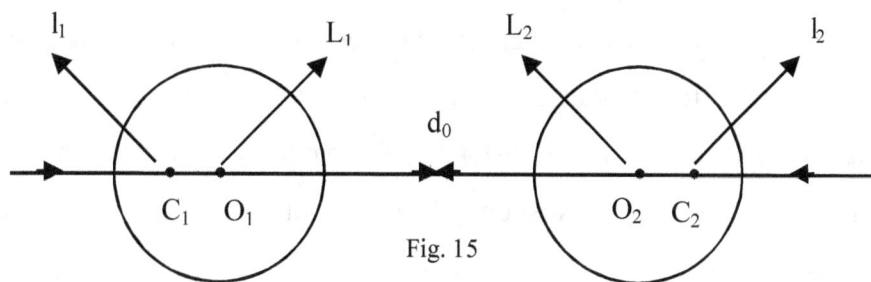

Fig. 15

Evolution of Physical Laws

Suppose r_1 and r_2 are the effective radii of the spheres of tranquillity of the Akashons (1) and (2). The lengths of separation of the mass and geometric centers of the two Akashons are l_1 and l_2 respectively.

O_1, O_2, C_1 and C_2 are geometric centers and the mass centers of the Akashons (1) and (2) respectively. The forces F_1 and F_2 acting on the two Akashons as a result of mutual interaction are given below.

$F_1 = F_{02}.l_1 / r_1, \quad F_2 = F_{01}.l_2 / r_2$

It is evident that as the distance of separation between the Akashons decreases, the extent of mutual interaction increases which leads to more attractive force between the Akashons. The unbalanced force actimg on the given Akashon depends on the force factor F_{02} or F_{01}, the displacement of the mass center l_1 or l_2 and the radius of the sphere of tranquillity r_1 or r_2. Out of the three factors the radius of the sphere of tranquillity decreases with decrease in the distance of separation due to creation of the extra region of perturbation but change is so small that it can be neglected, therefore, the force F_1 depends on two factors F_{02} and l_1. As already pointed out, the force is expected to increase with decrease of the distance, which leads to the conclusion that the force factor and the displacement should increase with decrease in the distance of separation. The force factor F_{02} which is associated with Akashon (2) has got different values at different points in the region of perturbation of interacting Akashon (2). The principle of inverse variation is applicable to find out its value at any point. The value of F_{02} is to be determined at point C_1. The distance of C_1 which lies in the region of perturbation of O_2, the geometric centre of Akashon (2) is $(r_1+r_2+d_0+l_1)$ therefore, according to the principle of inverse variation, the value of F_{02} at C_1 is given by the following relation.

$4\pi.(r_1 + r_2 + d_0 + l_1).F_{02} = K_E$ = universal constant

Similarly the value of F_{01} associated with the Akashon (1) can be deduced

$4\pi.(r_1 + r_2 + d_0 + l_1).F_{01} = K_E$

l_1 and l_2 being very small can be neglected

$F_{02} = F_{01} = k_E / 4\pi.(r_1 + r_2 + d_0) = k_E / 4\pi d$

Put $(r_1 + r_2 + d_0) = d$

$F_1 = (k_E / 4\pi d).(l_1 / r_1)$ and $F_2 = (k_E / 4\pi d).(l_2 / r_2)$

The above discussion shows that the force factor depends on the distance between interacting Akashons.

Suppose two Akashons having equal masses interact, the displacements of the mass centers of the two Akashons will be same, the radii of the spheres of tranquillity are also equal, therefore, the force $(k_E / 4\pi d).l$ acting on each Akashon will be same therefore $F_1 = F_2$. The above rule is applicable even to the Akashons with unequal masses. The result shows that action and reaction are equal and opposite. If it were not so the system will have some resultant force and hence motion.

The above result leads to the following conclusion

$l_1 / r_1 = l_2 / r_2, \quad l_1 / l_2 = r_1 / r_2 = m_2 / m_1$

m_1 and m_2 are masses of the mass center and geometric center of Akashons (1) and (2) respectively

$l_1 \alpha m_2$ or $l_1 \alpha 1 / m_1$

$l_2 \alpha m_1$ or $l_2 \alpha 1 / m_2$

The above result can be taken as such also without assuming the equal forces. As already seen the distance of separation of the Akashon is independent of its own mass when it interacts with the other Akashon. The above observation cancels the second option

$l_1 \alpha m_2$ $l_1 = k_2 m_2$

$l_2 \alpha m_1$ $l_2 = k_1 m_1$

As already discussed the decrease in distance leads to more mutual interaction as a consequence of which the distance of separation l_1 or l_2 should increase, therefore, the proportionality or displacement factors k_1 and k_2 like the factor F_{01} and F_{02} are expected to depend on the distance between interacting Akashons. Similar to the force factor, the values of displacement factors k_1 and k_2 are also given by principle of inverse variation

$4\pi . (r_1 + r_2 + d_0 + l_1).k_2 = k_0 =$ universal constant

$4\pi . (r_1 + r_2 + d_0 + l_2).k_1 = k_0$

neglecting the value l_1 and l_2

$4\pi . (r_1 + r_2 + d_0).k_2 = k_0$ $k_2 = k_0 / 4\pi d$

$4\pi \cdot (r_1 + r_2 + d_0) \cdot k_1 = k_0$ $k_1 = k_0 / 4\pi d$

$l_1 = (k_0 / 4\pi d) \cdot m_2$ $l_2 = (k_0 / 4\pi d) \cdot m_1$

$F_1 = F_{02} \cdot (l_1 / r_1)$ putting the values of F_{02} and l_1

$F_1 = K_E \cdot (k_0 / 16\pi^2 d^2) \cdot (m_2 / r_1)$

$r_1 = h / 4\pi m_1 \cdot c$

$F_1 = \{k_E \cdot (k_0 / 4\pi h) \cdot c\} \cdot (m_1 \cdot m_2 / d^2) = G(m_1 \cdot m_2 / d^2)$

$G = k_E k_0 c / 4\pi h$

Similarly, $F_2 = \{k_E \cdot (k_0 / 4\pi h) \cdot c\} \cdot (m_1 \cdot m_2 / d^2) = G \cdot (m_1 \cdot m_2 / d^2)$

$F_1 = F_2 = G \cdot (m_1 \cdot m_2 / d^2)$

The derivation of gravitational force is based on Hooke's concept of force.

There has been no comprehensive proof of the Newton's Law of Gravitation so far. The particle exchange concept of field has failed to derive the above equation.

The value of G points towards the existence of unique akashon N.

$G = K_E k_0 / 4\pi h$

The dimensions of K_E and k_0 are that of energy and $(16\pi^2 L^2)/M^2$

Suppose there is a unique akashon N having mass equal to M_0, K_E = constant energy = $M_0 C^2$

The radius of sphere of tranquillity of N akashon = $h/4\pi M_0 C$ = Constant = $(L)k_0 = (16\pi^2 L^2/\text{Mass})$

Put mass equal to M_0 and value of L equal to radius of sphere of tranquillity of N akashon, $k_0 = h^2/M_0^3 C^2$

Putting the values of K_E and k_0 in the expression for G it can be easily proved that $M_0 = \sqrt{hc/4\pi G}$ the relation gives mass of N akashon. The same relation has been derived independently later on also. The value of G is the value on the surface of sphere of tranquillity of N akashon.

The bulk form of matter can be considered to be equivalent to single akashon having mass equal to mass of cluster of akashons of which lump form of matter is composed of or because the bulk matter is cluster of akashons the gravitational force is the resultant of gravitational interaction of the akashons of which the bulk of matter is composed of.

EQUIVALENCE OF INERTIAL MASS AND GRAVITATIONAL MASS:

What is manifested as gravitational force is actually the unbalanced cosmic force. The gravitational interactions between two Akashons justify the shell structure of Akashon and presence of cosmic force. The full gravitational force experienced by the interacting Akashons is not instantaneous as in case of inertial force. The full gravitational pull is experienced only after the inertial time interval. The cessation of gravitational pull also needs this much time. **The m_1 and m_2 are not any special masses, the mass involved in the gravitational force is the same as used in the inertial force relation. It is quite evident that inertial mass is equal to gravitational mass. Therefore, gravitational force is inertial force and not geometrical force.** All the complicated precision experimental work to prove the equality of inertial and gravitational masses is mere wastage of time.

The matter in bulk form is composed of very very large number of Akashons. When the matter in bulk form experiences gravitational inter action, it is the attractive force experienced by the large number of Akashons present in the bulk form of matter. The resultant attractive force has got a force center. The bulk form of the mass can be considered to be equivalent to Akashon in which the center of the Akashon overlaps the force center of the bulk form and the mass of the single special Akashon is equal to the bulk form mass.

Evolution of Physical Laws

SPECIAL CASE OF INTERACTION:

The mutual interaction of the Akashons increases with the decrease of distance and the mass center is pushed more and more away as the distance decreases and at a certain stage the mass center just touch the surface of the sphere of tranquility. Let us discuss the case of mutual interaction of two special Akashons in which the two special Akashons are brought so close that surfaces of spheres of tranquillity of the two Akashons just touch and this is also the stage of minimum approach and the special Akashons are such that just as the surface of the spheres of tranquillity touch each other, the mass centers of the two Akashons also touch the surface of spheres of tranquility therefore, Akashon experiences maximum force. In this special interaction of special Akashons the value $d = 0$, therefore, the displacement of mass center l_1 or $l_2 = r_0 =$ the radius of the sphere of tranquillity of the special Akashon. The general relations associated with the displacement of the mass center are reduced to the following special form. m_0 is the mass of special Akashon during the special interaction, the mass centers of both the special Akashons are shifted to the maximum extent. The following displacement equation is applicable, k is displacement factor and r_0 is the displacement.

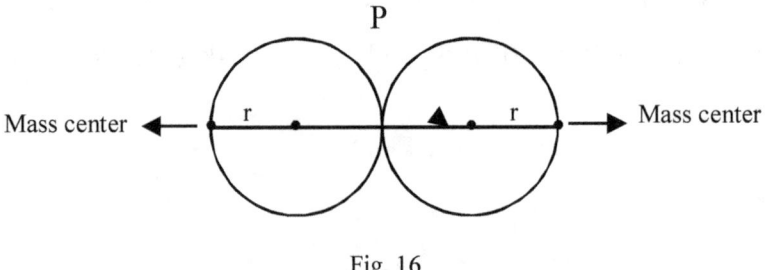

Fig. 16

Evolution of Physical Laws

The mass center lies on the surface of the sphere of tranquillity, therefore the following relations are applicable:

$4\pi.(4r_0).k = k_0,$ $\qquad r_0 = k.m_0,$ $\qquad k = r_0/m_0$

$k_0 = 16\pi.(r_0^2/m_0)$

In the special type of interaction, $K_E = 4\pi d.F_0$

F_0 is the maximum force, $F_0 = \pi h/16ct_0^2$

$d = 4r_0$

therefore, $K_E = 16\pi r_0.(\pi h/16ct_0^2)$

As already discussed in case of unique Akashon

$t_0 = 2\pi r_0/c$

Putting the value of t_0

$K_E = hc/4r_0$

$G_0 = K_E.k_0.c/4\pi h$

Put the value of K_E and k_0

$G_0 = (r_0/m_0).c^2 = k.c^2$

$G_0/r_0 = $ constant,

$G_0/r_0 = c^2/m_0 = $ constant, if r varies G also varies therefore,

$4\pi r G_0^{-1} = $ constant

It can be said that 1/G follows the principle of in verse variation. The conventional value of G is the value on the surface of sphere of tranquillity of N Akashon haqving r_0 radius which is constant hence G_0 is constant.

In the above expression, r_0 and m_0 are the radius and mass of unique Akashon and hence are constants.

The Change in Gravitational Constant G:

K_E and k_0 are properties of region of perturbation, therefore, the values depends upon distance from the mass center. G value depends on K_E and k_0, therefore, G is not constant everywhere or the value of G follows the principle of inverse variation.

$4\pi rG = k_G$

G is the value at a point on the sphere of tranquillity at distance r from the mass center of Akashon.

$dG/dr = - k_G/4\pi r^2$

Since k_G is very very less, therefore, when r is large $dG/dr \approx 0$ or it can be said that at large distance form the mass center, gravitational constant has got approximately constant value. But when r is very very small, dG/dr is not zero. Therefore, at points situated near the surface of sphere of tranquillity G changes with distance. Also value of K_E for heavier Akashon is more than that of lighter Akashons because $4\pi r.K_E = k_E$. It can be said that when the distance between the mass center is very very less, the heavier Akashon attracts the lighter Akashon with more force or action and reaction law breaks down. When the distance between the interacting Akashons is very very less of the order of radius of sphere of tranquillity, the value $G_0 = k_G/4\pi r$. Put the value of G in general gravitational force,

$F = k.(m_1.m_2/r^3)$ k is constant.

The value of G_0 at the surface of sphere of tranquillity of stationary Akashon as given by the principle of inverse variation,

$4\pi r_0 G_0 = k_G$

Similarly for moving Akashon,

$4\pi r G = k_G$

Therefore,

$G = G_0 r_0/r = G_0/[\sqrt{\{1-(v^2/c^2)\}}]$

THE PROOF THAT THE REGION OF PERTURBATION CONTAINS OSCILLATING SHELLS:

When two anti-particles (electron, positron) are brought near to each other the anti-particles change into two photons because the shells situated at the same distance from the center of the particles are out of phase. Therefore, annihilation of the two particles take place which shows the presence of oscillating region of perturbation.

THE GRAVITATIONAL CONSTANT AND SUPER HEAVY AKASHON(N):

According to the Newton's law of gravitation, gravitational constant $G = k_E \cdot k_0 \cdot c / 4\pi h$. As already seen $k_E = 4\pi \cdot (\text{Length} \times \text{Force})$.

Length × Force = Energy

Therefore, according to dimensions $k_E = 4\pi \cdot \text{Energy}$

Energy can be put equal to energy of super heavy or unique Akashon having mass = m_0. Therefore, energy = $m_0 \cdot c^2$

or $k_E = 4\pi m_0 \cdot c^2$

The value of k_0 as deduced earlier = $16\pi \cdot (r_0^2 / m_0)$

r_0 is the radius of sphere of tranquillity of special super heavy Akashon having mass m_0

Putting the value of k_E and k_0 in the relation for gravitational constant,

$G = 4\pi c^3 \cdot (r_0^2 / h)$

m_0 is the mass and r_0 is the radius of sphere of tranquillity.

Therefore, $4\pi r_0 m_0 = h/c$

Putting the value of r_0 into the equation for gravitational constant G and simplifying,

$m_0 = \sqrt{hc/4\pi G} = 1.539033 \times 10^{-8} \text{kg} = 8.6314 \cdot \text{ZeV} = 8.6314 \times 10^{12} \cdot \text{GeV}$

Evolution of Physical Laws

So it is seen that super Akashon is very very massive Akashon. Such heavy Akashons might have been created during the initial stages of creation and it is also possible that black holes are teeming with such Akashons. The super heavey Akashon is named as N elementary particle. The radius of sphere of tranquillity r_0 of super heavy Akashon or M_N,

$r_0 = h/4\pi mc = 1.1426 \times 10^{-35}$ m

The above radius value is very very small. It can be said that no length in nature can be measured less than this. This length can be taken equal to the radius of the vibrating shells of which all the Akashons are composed of. The dark matter and dark energy can also be considered of as huge clusters of these shells in stretched state but without any oscillations. Due to the stretched state, these shells can exert force but mass property is missing because no oscillations are there. The space is full of such shells having very very tiny radii.

The above relation can be deduced in another manner also. The two chosen Akashons are such that when the spheres of tranquillity of the two Akashons just touch each other the mass centers are also shifted away to just touch the surfaces of the spheres of tranquility, therefore, the Akashons experience maximum gravitational force. The distance d between the mass centers in the special case is $4r_0$; r_0 is the radius of spheres of tranquillity of unique Akashon.

$F_0 = Gm_0^2 / 16r_0^2$,

$F_0 = \pi \cdot h / 16 \cdot ct_0^2$

Let us find the value of t_0,

When the Akashon is subjected to maximum force, the mass centre of the Akashon just touches the surface of sphere of tranquility, this is the maximum displacement. When the oscillating mass centre of the Akashon just touches the sphere of tranquility the oscillating ether shell should be in mean position so that oscillations of the two centres may synchronize. This is possible if the time required by the Akashon centre to touch the surface of the sphere of tranquillity is equal to half the time period of the oscillating centre of the Akashon shell just present on the surface are equal. This time period can also be called inertial time period of the Akashon.

$4\pi r_0 v_0 = c$

$v_0 = c/4\pi r_0$

$t_0' = 1/v_0 = 4\pi r_0/c$

$t_0 = t_0'/2 = 2\pi r_0/c$

$G = (\pi.h/16.c.t_0^2).(r_0^2/m_0^2)$

Putting the value of t_0^2

$G = (r_0/m_0).c^2 = k.c^2$

$G/r_0 = $ constant

if r_0 and G both vary then $4\pi r G^{-1} = $ constant, G^{-1} follows the principle of inverse variation.

As explained earlier during the manifestation of matter the shells of the Akashon are squeezed inwards and are arranged around the sphere of tranquility, the above arrangement of the shells is not associated with the mass because of the lack of independent oscillations of the shells but the cluster in the above state can still experience force like the gravitational force on mutual interaction. The above state during the creation of matter can be called dark matter. The dark matter unlike the conventional matter is not associated with the independent oscillations of the shell but due to inwards squeezing of the shells, the manifestation of the secondary force on mutual interaction is possible. The dark matter can be called as half-baked cookie in the cosmic confectionary shop. The proper Akashon is born when the displaced shells arranged around the sphere of tranquillity are imparted independent oscillations and with the special nature of the shells the Akashon or Prakashon gets the property of exchange of matter or energy on mutual interaction. The above property is missing in the dark matter because in this stage the shells are merely displaced inwards with Cosmic Force without any independent oscillations. The displaced shells exhibit gravitational effect. The neutrino can travel very very long distance through the thick layers of conventional matter because the neutrino contains displaced shells without any oscillations and hence it lacks the property of interaction. In fact the half-baked state of matter is more abundant than the fully baked conventional matter as if there were shortage of fuel during the cooking of cosmic beans.

As already mentioned the black holes are eating up the conventional matter and at certain stage when the stomach is full, the black holes might be discarding the black matter as Cosmic refuse after assimilating the independent oscillations of the shells of the Akashons of the conventional matter. The black holes are the plants where manifested state is being continuously changed into un-manifested state. Let us next find out the mass of the unique or NAkashon.

SUPER HEAVY AKASHON OR ETHERON (OR N AKASHON):

Let us again take the case of two special Akashon interaction. The Akashons are such that when the spheres of tranquility of the Akashon just touch each other the mass centres of the two Akashons just touch the surface of the sphere of tranquility (see the fig.11). At this stage, the Akashons experience maximum gravitational force, the distance between the mass centres is $4r_0$.

Gravitational force relation is valid even up to very small distance when interacting Akashon are pure mass because then only gravimetric interaction is manifested.

The maximum gravitational force = $G.m_0^2 / 16r_0^2$
The maximum inertial force to which Akashon can be subjected is $hc/16\pi t_0^2$
$t_0 = 2\pi r_0/c$ (this relation has already been derived)

Putting the value of t_0 in force relation,
Maximum inertial force = $hc / 64\pi r_0^2$

Equating the gravitation force and maximum force
$G.m_0^2 / 16r_0^2 = hc / 64\pi r_0^2$
$m_0 = \sqrt{(hc / 4\pi G)} = 1.539033 \times 10^{-8}$ kg

m_0 is the maximum pure mass of N Akashon. In general pure mass can be expressed as $m = \sqrt{(hc / 4\pi G)}$. Where, G is the value on the surface of sphere of tranquility. The pure mass can be expressed in the same units as that of hybrid or conventional mass.

Pure mass $m_p = hc/4\pi G$

$4\pi r_p G = k_G$

For hybrid mass m, $4\pi rm = h/c$

Using the above three equations, if

$m_p = m$, the following relation between the radii of pure mass and hybrid mass is obtained.

$r_p . r^2 = hk_G/16\pi^2 c^3 = $ constant

the above relation shows that by making appropriate changes in the radii, the value of pure mass can change to magnitude of hybrid mass and vice versa.

The above result can be got by using dimensions also

$L.(T)^{-1} = c$

$(L^3/M).(T)^{-2} = G$

squaring first relation, $L^2 T^{-2} = C^2$, dividing by G relation,

$L/M = G/c$

$L = 4\pi Mc$

$L = h/4\pi mc$

Putting the value of L and rearranging,

$$m_0 = \sqrt{(hc/4\pi G)}$$

here, M the mass dimension is replaced by m_0 the N Akashon.

The above elementary particle can be produced in very high energy conditions. It is unique in nature. The radius of sphere of tranquility of the unique Akashon $r_0 = 1.1426 \times 10^{-35}$m. The unique N Akashon is the heaviest elementary particle. The radius of the sphere of tranquillity of this Akashon can be put equal to the radii of the shells of which all the Akashons are composed of or the value of this radius is equal to the radii of the bubbles of which the space froth is composed of.

GENERAL THEORY OF RELATIVITY:

The knowledge of the gravitational field and the model of the Akashon can be used to determine certain results in very simple manner without using any complicated mathematical calculations involving tensor algebra. The cornerstone of general theory of relativity is Einstein's principle of equivalence. One of the statements of the principle of equivalence expresses that in the neighborhood of any given point, it is not possible to distinguish between the gravitational field produced by the attraction of mass and the field produced by the acceleration of the inertial frame of reference. It can be proved very easily by using the model of Akashon that why an accelerated Akashon behaves like an Akashon placed in the gravitational field because displacement of mass center is the same when the Akashon is accelerated or placed in the gravitational field. It can be seen that **the principle of equivalence is no longer matter of experience, it has theoretical justifications also.** As seen earlier in the formula for the Newton's force relation and the gravitational force relation, same value of mass is used. According to the principle of equivalence, the inertial and the gravitational effects are equivalent. As already seen while discussing the concept of force, the mass center and geometric center of an accelerated Akashon are separated when inertial force is applied, the direction of shifting away of the mass center is opposite to the direction of force. The same thing happens when two Akashons experiences gravitational force during mutual interaction. The separation of mass center and the geometric center is equivalent whether the separation is brought due to inertial time lag or it is the result of the mutual interaction of the two Akashons. The principle of equivalence is not merely matter of chance or experience. The solid foundation of the principle of equivalence is based on the force relation F α l/L, in both the cases when the Akashon is accelerated or it is experiencing gravitational force. There is no need to do so much useless theoretical or practical work to simply prove that inertial and gravitational masses are equivalent. The equivalence of gravitational and inertial masses needs no complicated justification because according to the line of treatment adopted here, the equivalence is self-evident and there is only one type of universal mass and the behaviour of that mass is same in all situations.

It has been seen earlier that radius of sphere of tranquillity when the Akashon is moving with uniform velocity v is given by

$$r = r_0 \sqrt{\{1-(v^2/c^2)\}}$$

it has also been calculated that if the Akashon is being attracted or accelerated by gravitating body the radius of sphere of tranquillity when the Akashon is at distance d from the central attracting force

$$r = r_0 \sqrt{\{1-(2GM/c^2d)\}}$$

The first equation as already discussed is the backbone of special theory of relativity while the second equation when the Akashon is accelerated is the corner stone of general theory of relativity.

While discussing general theory of relativity radius of the Akashon will be replaced by l while distance d by r because usually in general relativity distance is expressed as such.

The measurement of length and time in gravitational field:

Gravitational field Ψ at distance d from the gravitating body is given by

$\Psi = -GM/r$

In case of general theory of relativity

$l = l_0 \sqrt{\{1 - (2GM/c^2 r)\}} = l_0 \sqrt{\{1 + (2\Psi/c^2)\}}$

It is seen that radius of sphere of tranquillity decreases in gravitational field. The Aakashon in gravitational field can be used as standard meter and standard clock to measure distance and time in the same manner as in case of special theory of relativity. By the same arguments as in case special theory of relativity if L and L_0 are the distances measured in the presence and absence of gravitational fields

$L = L_0 \sqrt{\{1 - (2GM/c^2 r)\}}$

$L < L_0$

There is length contraction if elements of length dr and dr_0 are used, dr and dr_0 and are the very very small radii of spheres of tranquillity of special test akashons discussed later on in the absence and presence of gravitational field   The mass of the test akashon is less than half of the mass of N akashon.

$$dr = dr_0 \sqrt{\{1 - (2GM/c^2 r)\}}$$

Because dr_0 is very very small which shows that test akashon used is very massive. The presence of outside gravitational field only shortens the radius of sphere of tranquillity that is the space inside the miniscule volume of sphere of tranquillity. From objective point of view the outside space is not curved due to the presence of gravitational field. The akashon will move in straight path towards the gravitating body with the difference that its radius of sphere of tranquillity will change hence change in mass will lead to increase in gravitational force. The concept of geodesic is only bizarre physical interpretation of abstract mathematical equations. The path for the motion is not predestined by the gravitating body. It will be shown soon that Schwarzschild equation has been derived vey simply by using basic equation based on inertial force hence gravitational force is inertial force. It is not the result of curvature of space but the cause of gravitational force is stress of akasha .The advance of the perihelion of mercury and bending of light has been proved later on in this topic very simply without using tensor algebra and thus rending redundant the concept of curved space.

Similarly, as in case of special theory of relativity when Akashon is used as time measuring device the Akashon clock is slowed down. If t and t_0 are the rates of flow of time or times required to complete one oscillation by the shells of the Akashon on the surface of the sphere of tranquillity in the presence and absence of gravitational field.

Suppose T' and T_0' are the times required for n ticks of the Akashon clock in the presence and absence of gravitational field. As in the case of Akashon moving with uniform velocity the same relation is applicable here, t and t_0 are the time periods between ticks in absence and presence of gravitational field.

As in case of special theory of relativity $t \propto l$ and $t_0 \propto l_0$

Therefore, $T' \propto nl$ and $T_0 \propto nl_0$

$T'/T_0' \propto l/l_0 = \sqrt{\{1 - (2GM/c^2r)\}}$

Which shows that $T' < T_0'$ or clock is retarded in gravitational field.

Suppose T and T_0 are the times required for the completion of the same process by using Akashon clock in the presence and absence of gravitational field,

$T \propto 1/$(rate of flow of time)$\propto 1/$(time unit) $\propto 1/$(radius of sphere of tranquillity)

As discussed in special theoty of relativity,

Thus, $T \propto 1/t \propto 1/l$

$T_0 \propto 1/t_0 \propto 1/l_0$

Thus, $T/T_0 = l_0/l = 1/[\sqrt{\{1 - (2GM/c^2r)\}}]$

$T > T_0$, which shows that clock is retarded in gravitational field.

For small intervals $dt_0/dt = \sqrt{\{1 - (2GM/c^2r)\}}$

Suppose m is the relativistic mass of the Akashon gravitational field and m_0 is the fixed mass,

$m/m_0 = l_0/l = 1/\sqrt{\{1 - (2GM/c^2r)\}}$

$m > m_0$

If the effect of velocity v is taken into account in all the above expressions factor $\sqrt{\{1 - (2GM/c^2r)\}}$ will be replaced by $\sqrt{1 - [(2GM/c^2r)\{1 - (v^2/c^2)\}]}$, which approximately equal to $\sqrt{\{1 + (2\Psi/c^2) - (v^2/c^2)\}}$. The new factor accounts for the contraction due to velocity.

Minkowski Space:

The equation $4\pi r/t = c$ can be called as concise form of Minkowiski Space. Suppose center of reference Akashon overlaps the origin of a coordinate system, point P(x,y,z) lies on the surface of sphere of tranquillity having radius equal to r.

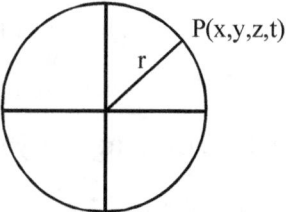

Reference Akashon

Fig. 17

Evolution of Physical Laws

$x^2 + y^2 + z^2 = r^2$

$4\pi r/t = c$

Putting the value of r^2

$16\pi^2(x^2 + y^2 + z^2) - c^2t^2 = 0$

$(x^2 + y^2 + z^2) - (c^2t^2/16\pi^2) = 0$

The above equation shows that $(x^2 + y^2 + z^2) - c^2t^2$ can be negative, positive and zero.

Using positive value $s^2 = c^2t^2 - (x^2 + y^2 + z^2)$

$-s^2 = (ict)^2 + (x^2 + y^2 + z^2)$

In the above equation, along with the squares of usual coordinates (x,y,z) square of fourth imagenary term *ict* involving time is added which hints at considering time as the fourth cooridinate. As (x,y,z) of a point P change with the change in reference Akashon t also changes which associate time with space coordinates x, y, and z. Actually (x,y,z,t) are associated as such according to the special relation $4\pi r/t = c$ as given by the principle of inverse variation. The starting equation $4\pi r/t = c$ is 3-dimensional and hence equation got by using 3 dimensions can lead to only fictitious space involving time as fourth dimension. This fictitious space is called as Minkowiski space. Each point of Minkowiski space can be represented by a point lying on the surface of sphere of tranquillity of reference Akashon.

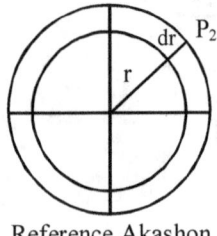

Reference Akashon

Fig. 18

Differential Minkowiski equation is got by considering two points P_1 and P_2 lying on the surfaces of concentric spheres of tranquillity having radius equal to r and (r+dr)

$$ds^2 = c^2 dt^2 - (dx^2 + dy^2 + dz^2)$$

In polar coordinates,

$$ds^2 = c^2 dt^2 - dr^2 - r^2(d\theta^2 + \sin^2\theta \cdot d\phi^2)$$

dt is the difference in time intervals for the completion of the same process performed at P_1 and P_2.

The above equation in the absence of gravitational field is written as,

$$ds_0^2 = c^2 dt_0^2 - dr_0^2 - r^2(d\theta^2 + \sin^2\theta \cdot d\phi^2)$$

Schwarzschilds equation:

It is alleged that it is impossible to get simple solution of Schwarzschilds metric. The solution presented here is not only simple but it is the simplest possible solution. The Schwarzschilds equation is simply Minkowiski equation in the presence of gravitational field. Suppose length and time element change to dr and dt in the presence of gravitational field.

$$ds^2 = c^2 dt^2 - dr^2 - r^2(d\theta^2 + \sin^2\theta \cdot d\phi^2)$$

Minkowiski equation in the absence of gravitational field is

$$ds_0^2 = c^2 dt_0^2 - dr_0^2 - r^2(d\theta^2 + \sin^2\theta \cdot d\phi^2)$$

As already discussed $dr = dr_0 \sqrt{\{1 - (2GM/c^2 r)\}}$

Similarly,

$dt_0 = dt \sqrt{\{1 - (2GM/c^2 r)\}}$

Putting the values of dr_0 and dt_0 in the Minkowiski equation in the absence of gravitational field the equation changes to new form in the presence of gravitational field, due to gravitational field of mass M placed at the centre of reference akashon sphere of tranquillity there is change in the length and time elements as deduced earlier without using tensors

$$ds^2 = \{c^2dt^2(1 - 2GM/c^2r) - dr^2/(1 - 2GM/c^2r)\} - r^2(d\theta^2 + \sin^2\theta \cdot d\phi^2)$$

The above equation is called Schawrszchild equation which has been derived by using very simple mathematics and without using Einstien's equation and equivalent principle which need very complicated tensor algebra to get Swarchzchilds equation as a solution of Einstien's equation. The Schwarzschild equation has been derived very easily without using bizzare concept of curved or 4-dimensional space. As aready discussed radius of sphere of tranquillity l and l_0 in the presence and absence of gravitational field are given by

$$l = l_0 \sqrt{\{1 - (2GM/c^2r)\}}$$
$$m/m_0 = l_0/l = 1/\sqrt{\{1 - (2GM/c^2r)\}}$$

The equation can be used for calculating advance of perihelion of mercury and deflection of light.

The above equation shows that when $(2GM/c^2r) = 1$ or $r = c^2/2GM$ i.e. when the Akashon reaches distance r given by the above equation the relativistic weight of m of Akashon will approach infinity so distance r gives a point of singularity. So long as $2GM/c^2r < 1$, mass is finite or when $r > 2GM/c^2$, no singularity is observed.

$r = 2GM/c^2$ is called Schwarzschilds radius.

The above equation shows that when r is equal to Schwarzschilds radius or when the Akashon is at a distance equal to Schwarszchilds radius from the gravitating mass, test Akashon should change to blackhole but it does not happen practically because at a certain distance relative mass of the Akashon will become equal to M_N Akashon which is unstable. Hence, mass of the Akashon will not further increase due to gravitational interaction and no singularity will be observed. The relativistic mass of an akashon in gravitational field is given $m = m_0 / \sqrt{1 - 2GM/c^2 r}$. The equation shows that when $r = 2GM/c^2$ relativistic mass m should become infinity or point of singularity is reached but practically singularity is not achieved because before reaching distance equal to Schwarzschild radius the test mass gathers mass equal to N Akashon which is unstable hence further increase in mass is stopped without achieving state of singularity. If the test mass is electron it will break up into electron and highly energetic photon when relativistic mass of electron becomes equal to N akashon The distance at which relativistic mass of electron becomes equal to N akashon is calculated below

m = mass of N akashon = $\sqrt{hc}/4\pi G = 1/\sqrt{1 - 2GM/c^2 r}$.

The value of r when relativistic mass becomes equal to N akashon

$R = 2GM/c^2 \{1/(1-4\pi G\, m_0^2/hc\,)\}$

The above equation proves that $r > 2GM/c^2$

In other words the test akashon changes to unstable N akashopn state before reaching point of singularity or distance equal to Schwarzschild radius. The mass of an akashon or conventional mass packet composed of doublet of hybrid mass-charge shells having shell radius equal to radius of sphere of tranquillity of N akashon cannot be greater than N akashon. The black hole is composed of hybrid quadruplet or mayons of four shells, two of mass shells (akasha) and two of chargeshells (antiakasha). The radius of hybrid quadruplet shell is much smaller than that of hybrid doublet shells hence the minimum radius of sphere of tranquillity of akashon (black hole) composed of hybrid quadruplet is very very small hence mass of black hole is very huge but even then state of singularity is not reached because though radius of sphere of tranquillity is very less but it is not zero. Derivation of Schewarzschilds equation is derived while retaining the three dimensional plane space. The gravitational force is just like the inertial force and it is not the result of curvature of space.

Equation for the motion of photon or light in the four dimensional space named as N space:

Minkowski space is for the motion of mass particles and as already seen is got by taking positive value of s.

If s=0 or ds=0 another type of space called N-space is got for light

$0 = c^2 t^2 - (x^2+y^2+z^2)$ or $o = c^2 dt^2 - (dx^2+dy^2+dz^2)$

The above equation shows that points of N space are located on the surface of reference sphere of tranquillity of photon having radius equal ct where t is the timed required for light starting from centre of reference sphere of tranquillity to the surface. If mass M is placed at the centre of sphere, due to gravity of mass there are changes in time and length element deduced earlier without using tensors therefore N space changes to special equation called N equation which is applicable to the motion of light or photon. N equation is given below.

$$0 = (1-2GM/c^2r)c^2 dt^2 - dr^2/(1 - 2GM/c^2r) = (1-2GM/c^2r)c^2 dt^2 - (1+2GM/c^2r) dr^2$$

Einstein got the following equation by using tensors
$$ds^2 = (1-2GM/c^2r) c^2 dt^2 - (1+2GM/c^2r) dr^2$$

He arbitrarily put ds=0 to get equation similar to N equation got earlier without using any tensor algebra. N equation can be used to calculate angle of deflection of light in gravity. Velocity of light in gravitational field v,

$$v = dr/dt = c\{\sqrt{1 - 2GM/c^2r}/\sqrt{1 + 2GM/c^2 r}\}$$

The above equation shows that speed of light becomes less as if there is refraction effect. Einstein used Huyghens principle to calculate angle of deflection. If light is moving in x direction z=0 plane the transverse deflection α is as under

Evolution of Physical Laws

$d\alpha/dx = 1/c(dy/dx) = 2GMy/c^2 dr^3 = 2GM/c^2 \{y/x^2+y^2\}$

on integration w.r.t x $\alpha = 4GM/c^2 R$

Evidences in favour of results of general theory of relativity:

1. *Gravitational red shift:* Suppose l and l_0 are radii of spheres of tranquilillity of photon in the absence and presence of gravitational field respectively, λ_0 and λ are the respective wavelengths of the photon.

$\lambda_0 = 4\pi l_0$

$\lambda = 4\pi l$

$\lambda_0/\lambda = l_0/l = \sqrt{\{1 - (2GM/c^2 r)\}}$

$\lambda = \lambda_0 \sqrt{\{1 - (2GM/c^2 r)\}}$

The above equation shows that $\lambda < \lambda_0$ or wavelength in the presence of gravitational field is less. As r increases λ also increases which shows that as photon moves away from the gravitating mass wavelength becomes more and more or it is shifted to red part of the spectrum.

2. *Deflection of light:* The deflection of light in gravitational field has already been discussed without using tensors

3. *Advance of the peihelion of the planet mercury:* Usually Schwarzschilds equation derived above is used to calculate the above result. Here the shift in Perihelion can be calculated by converting the

classical equation of motion of the planet into relativistic equation which is very easy and does not need any tensor algebra etc. Advance of the perihelion has been calculated by using relativistic mass of the planet in gravitational field and Newton's gravitational force law.

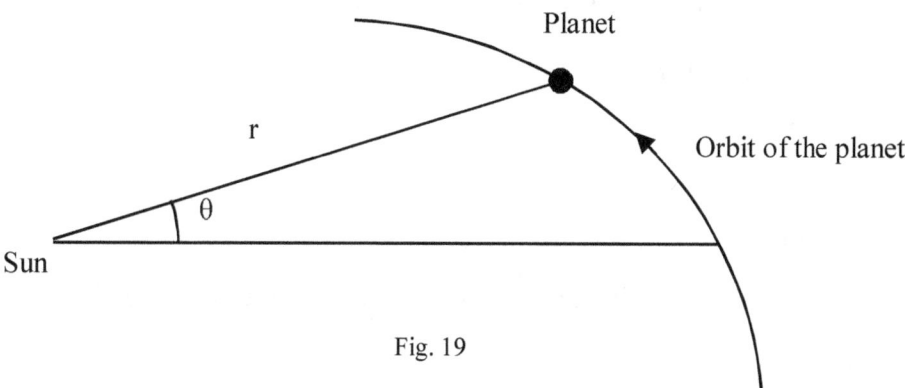

Fig. 19

The classical equation of motion of planet is

$d^2u/d\theta^2 + u = -(\mu/l^2u^2)F$

$u = 1/r$, μ is reduced mass, F is the gravitational force acting on the planet, and $l = \mu r^2 \cdot d\theta/dt$

$r^2 \cdot (d\theta/dt) = h$

The above equation is reduced to the following form

$d^2u/d\theta^2 + u = -(\mu/h^2u^2).F$

When the classical equation is to be changed to relativistic form L.H.S. will not change because u depends on r, which is very large and does not undergo appreciable contraction.

Suppose, $d\theta_0$ and $d\theta$ are the values in the absence and presence of gravitational field,

$d\theta_0 = dl_0/r$

$d\theta = dl/r$

$dl = dl_0 \sqrt{\{1 - (2GM/c^2r)\}} \approx dl_0\{1 - (GM/c^2r)\}$

$dl/r = (dl_0/r) - (GM.dl/r^2)$

r is very large and dl is very small.

$d\theta \approx d\theta_0$

Thus there is no change in L.H.S. The change in the R.H.S. of the equation,

Force F in the presence of gravitational field undergoes change because relativistic mass m of the planet increases.

$m = m_0/\sqrt{\{1 - (2GM/c^2r)\}}$

The change in the mass M of the sun can be neglected because M is very large.

Force F in the presence of gravitational field,

$F = GmM/r^2 = [m_0/\sqrt{\{1 - (2GM/c^2r)\}}].M/r^2$

Change in the value of h,

Suppose h and h_0 are values in the presence and absence of gravitational field,

$h = r^2 d\theta/dt$

$h_0 = r^2 . d\theta/dt_0$

$d\theta$ is nearly equal to $d\theta_0$

$d\theta/dt = (d\theta/dt_0).(dt_0/dt) = (d\theta/dt_0).[1/\sqrt{\{1 - (2GM/c^2r)\}}]$

The above relation shows $h = h_0/.[1/\sqrt{\{1 - (2GM/c^2r)\}}]$

Putting the relativistic values of h and force, and taking reduced mass equal to unchanged value m_0 due to very large value of mass of the sun. Reduced mass can be taken equal to non-relativistic mass value of planet because change in mass while calculating reduced mass due to relativistic effect is very less.

The classical equation changes to the following form,

$d^2u/d\theta^2 + u = (MG/h^2).\{1 - (2GM/c^2r)\}^{-3/2}$

Neglecting expansion terms involving powers of c more than c^2

$d^2u/d\theta^2 + u = (MG/h^2).\{1 + (3GM/c^2r)\} = GM/h^2.(1 + 3kM/r)$

$G = kc^2$

The above equation can be changed into the conventional form. The centripetal force acting on the planet is equal to the gravitational force.

$mr(d\theta/dt)^2 = GmM/r^2$

$(d\theta/dt)^2 = GM/r^3$

$r^4(d\theta/dt)^2 = GMr$

$h = r^2(d\theta/dt)$

$h^2 = r^4(d\theta/dt)^2$

$h^2 = GMr = kc^2Mr$

The equation of motion can be changed to the new from as given below

$d^2u/d\theta^2 + u = (GM/h^2) + (3GM^2k/rh^2)$

as already seen $G = kc^2$, $h^2 = kc^2Mr$, $u^2r^2 = 1$

$d^2u/d\theta^2 + u = (GM/h^2) + (3GMu^2/c^2)$

The above approximate equation of motion is of the same form as derived by using tensor algebra. The tensor based equation gives only one extra term on the right hand side as compared to the Newtonian equation but more terms can be added to the right hand side of the terms involved in the equation without using tensor algebra if terms are not neglected. The equation involving the third term is given below.

$d^2u/d\theta^2 + u = (GM/h^2) + (3GMu^2/c^2) + (6GMu^3/c^4)$

So far we have seen that the cause of the manifestation of the gravitational field is not the unrealistic concept of the curved space but the manifestation of gravity as the unbalanced primary Cosmic Force.

As already seen, the force acting on the planet is given by the following relation

V is the potential energy of planet, $V = -Gm_0/[\sqrt{\{1 - (2GM/c^2r)\}}]$

$F = -dV/dr = (Gm_0M/r^2)[\{1/\sqrt{(1 - 2kM/r)}\} + \{kM/r(1 - 2kM/r)^{3/2}\}]$

If the factor kM/r which is very small is ignored, the force relation is reduced to the following form

$F = -dV/dr = (Gm_0M/r^2) + (2kGm_0M^2/r^3)$

The above relation shows that the two forces are acting on the planet. The magnitude of the first force follows inverse square law while the second force depends on the cube of the distance from the sun. the contribution of the second force is very small because it involves the factor $k = G/c^2$. The path traced by the planet (Akashon) under the influence of the first major force which varies as the inverse of the square of the distance is ecliptical while the shape of the orbit is spiral in shape under the influence of second force. The fusion of the two orbits results in the advance of the perihelion of the planet. Thus the cause of the perihelion is the comparatively second weak force. The gravitational interaction is not instantaneous because during this interaction, gravitational field is manifested only after the rearrangement of shells.

Deflection of light in gravitational field:

Angular momentum of Akashon is l,

$l = mr^2 \cdot d\theta/dt = m \cdot h$

$h = l/m$, if $m \to 0$, $h \to \infty$

Mass of photon = 0, therefore, mg/h_0^2 of the relative equation of motion of the planet is zero. Therefore, equation of motion of the photon in the presence of gravitational field is reduced to

$$d^2u/d\theta^2 = 3GMu^2/c^2$$

This is called null geo-desic equation. Both the relativistic equations of motion for the planet and the photon can be used to calculate advance of the Perihelion and deflection of light.

Exact classical value by classical mechanics of bending of light:

The eccentricity of path followed by mass m in gravitating body of mass M is as follow eccentricity = $(\sqrt{1 + 2L^2E/G^2m^3M^2})$. E and L are energy and angular momentum. Neglecting gravitational energy E = kineticenergy = $(1/2)mv^2$. The distance from gravitating body is R. As discussed earlier the energy of photon is shared by electromagneton and mattenerg in the presence of strong gravitational field energy is equally shared by electromagneton and mattenergon or it can be said that mattenergon carries mass = m/2 if m is the energy equivalent mass of photon. Gravitational field is associated with mattenergon only hence deflecting force is experienced by mass m/2 of photon, m is the mass equivalent of photon energy. L = (1/2)mvR, while calculating energy mass is taken as m but while calculating angular momentum half of mass of photon associated with the mattenergon is considered because only mattenergon experiences gravitational force. Therefore, angular momentum L=1/2 m vR. While calculating energy mass is not halved because mere bending of light does not affect the energy. Only the mattenergon experiences the gravitational force. The photon velocity c is very high therefore while calculating eccentricity factor 1 is neglected and approximate value of e is given below.

$e \approx c^2 R/2GM$

Angle of deflection $= \pi - 2\cos^{-1} 1/e$

The value of second factor is $\cos^{-1} 1/e = \pi/2 - 1/e$

Angle of deflection $= 2/e = 4GM/C^2 R$

The value of angle of deflection got by using classical mechanics is the same as got from the theory of general relativity where very higher mathematics is used. Moreover there is no physical explanation for the bending of light. The bending of light also supports the composition of photon.

Unification of general theory of relativity, quantum mechanics and quantum gravity: In thought experiments there is no limit to distance between two points fixed in space; it can be as small as nearly zero so space is said to continuous but practically it is not possible. In thought experiment two points can be positioned at imaginary points in space but in reality location of the positions of the points is not possible. As already discussed the diameter or radius of sphere of tranquillity of an akashon is the natural meter. Because end points of radius can be practically located in space. As already discussed N-akashon has got the smallest value of radius of sphere tranquilliy hence a_0 the radius of the sphere of tranquillity is the smallest unit of length, the practically measurement of length less than it is not possible though space length can be less than it also. The value of smallest length unit is of the order of 10^{-35}. The constraint on the practical measurement of minimum length quantizes the space because no metre rod is there smaller than this value. The value of minimum space that can be measured is so small that practically continuity of space is not much affected hence the results of general theory of relativity are not much off the mark. The contraction in the radius of sphere of tranquillity of test akashon having mass less than or equal to half the mass of N akashon is calculated by placing it in the gravitational which is important result of general theory of relativity The contaction of length is associated with time dilation as already discussed in general theory of relativity. Why general relativity is applicable to akashon with mass less than half the mass of N Akashon but not more than that is discussed further in this topic. The quantum mechanics rules are also applicable to the test akashon which shows that both the theories are applicable at submicroscopic scale. It is erroneous to say that general theory of relativity is applicable to macroscopic state only though it will be discussed while examining quantum gravity related to this topic that general relativity at microscopic level is applicable only to isolated microscopic particles with mass less than half the mass of N akashon. There is no major conflict regarding the field of application of two theories It is the special nature of model of akashon that helps in removing the conflict. Quantum mechanics operates at macroscopic level also if the mass of isolated akashon is very high but the effects are negligible when measured

The proposed models of akashon and prakashon (photon) impart non local nature to both because both are associated with limitless regions of perturbation and hence have got limitless range of interaction .It is through limitless regions of perturbations that entanglement of photon and spooky at distance in akashons is possible .The proposed models of akashon and prakashon(photon}are such that centre can be located at a point inside the sphere of tranquillity when at the same time akashon and prakashon are having extended sizes also.. The proposed models of akashon and prakashon(photon) resolve the conflict related to non locality because both are non local in nature due to association of limitless regions of perturbation It can be said conflicts are there because of ignorance about the basic nature of fundamental units related to mass(akashon)and energy (photon) The mass centre of an akashon can be located anywhere inside the sphere of tranquillity having very very small volume. In case of electron volume of sphere of tranquillity is of the order of $10^{-39} m^3$ therefore error in locating the centre inside the sphere of tranquillity is so small that it can be neglecred. God plays dice but not with not so uncertainity as the mainstream quanthm mechanics proposes. All the energy results of microscopic particles can be exactly deduced by applying mattenergon quantum mechanics as discussed in the present book without introducing the concept of probability at all. Einstein space of relativity is not perfectly continuous but as seen above the quantum jumps are too small to introduce grave errors in the results of relativity. Gaps have to be introduced in space on account of imposition of constraint smallest limit of measureable length otherwise the space is smooth. Discontinuity is only artificial.

Quantum gravity: As already discussed the radius of sphere of tranquillity l and l_0 of an akashon in absence and presence of gravitational field are related as $l=l_0\sqrt{1-2GM/c^2 r}$. It is also known that radius of sphere of tranquillity is multiple of minimum length

$l = na_0$ and $l_0 = n_0 a_0$

n_0 = 1, 2, 3, ...

n = (n_0-1), (n_0-2) or n = 1, 2, 3, ...

0 value of n is not allowed

The relation is so that radius change takes by one unit of minimum length. When n_0=1 the minimum value of n=0 but this is not allowed because the zero radius of sphere of tranquillity of test akashon in gravitational field will lead to infinite mass. The allowed minimum value is n_0=2, the allowed radius of sphere of tranquillity of test akashon=$2a_0$, which is twice the value of radius of sphere of tranquillity of N akashon hence the maximum mass of test akashon used should be equal to less than half the mass of test akashon because mass of akashon is inversely proportional to radius of sphere of tranquillity. This point is discussed in following manner also.

$n = n_0 \sqrt{1 - 2GM/c^2 r}$

hence, $GM/r = c^2/2(1-n^2/n_0^2)$

The gravitational potential $V = -GM/r = -c^2/2(1-n^2/n_0^2)$

The above equation shows that as the test akashon moves away or towards the gravitating body the change in gravitational potential is not continuous but quantized because n = 1, 2, 3, ……. Thus gravity is quantized because there is stepwise change in potential.

At point 1 put $n=n_0-1$
Find out value of V_1
Similarly at point 2 put $n=n_0-2$ and find out value of V_2
$V_2-V_1=(c^2/2)(3-2n_0)/n_0^2$

The above equation shows that as test akashon moves away or towards gravitating body the change in gravitational potential takes place in steps of fixed magnitude. The potential difference at point 2 and at point 1 should be negative as test akashon is moved from point 1 to point 2 towards gravitating body of mass M on account of attractive nature of force, it is so only if $n_0>3/2$ but because n_0 is whole number therefore its value is expected to be greater than or equal to 2.

Therefore minimum value of radius of test akashon is $= 2a_0$

(mass of test akashon)/(mass of N akashon), $a_0/2a_0=1/2$

This relation is applicable because mass is inversely proportional to radius therefore mass of test akashon is half of mass of N AKASHON if mass is more than half of N akashon the law of gravity will breakdown for an isolated akashon, the upper limit of mass for the operation of gravity is equal to half the mass of N akashon. The mass of N akashon $= \sqrt{hc}/4\pi G =$ 8.63148×10^{21} Mev. When the matter is present in lump form the law of gravity operates though mass is much more than half of mass of N akashon. it is so because bulk matter is composed of very large number of akashons each having mass much less than the N akashon hence law of gravity is applicable to these akashons. The resultant gravity interaction is gravity interaction of the akashons of which the bulk of matter is composed. The resultant force due to interacton of akashons in bulk form acts at a point in the bulk of matter called as mass centre. The gravitational force acts along the line joining the mass centres.

Suppose if the distance r joining the mass centres of lumps is quantized
$r = Na_0$
N is very huge number because r is very large and a_0 is very very small
$V = -GM/r = -GM/Na_0$

Due to huge value of N it can be easily shown by differentiating that V will change smoothly as the test akashon in lump form moves towards gravitating body or there is no quantization of gravity. Quantization is there only when the test akashon has got mass equal to or less than half the mass of N akashon.

Theories of modern physics seem to be discordant because .natural phenomena are being studied in isolated approach just like a group of blind persons exploring an elephant and drawing their own conclusions about the shape of elephant because no one has seen the elephant as a whole.

CALCULATION OF MASS OF ELEMENTARY PARTICLES:

The minimum measureable value of length a_0 is of the order of 10^{-35} m, therefore radius of sphere of tranquillity is expressed as $r = na_0$ because minimum length is very very small therefore n should very large for r to have appreciable value. The minimum value of n=1 therefore maximum value of mass is associated with N akashon = $h/4\pi a_0 c$

In general mass of any akashon is quantized $m = h/4\pi ncr_0$, n is an integer in ideal case, if it changes continuously, and somehow the following mass relations give good results but are not so logical.

Hence, on differentiating $dm/dn = \dfrac{h}{4\pi a_0 n^2} = -\dfrac{m}{n}$

On integrating both sides, $\ln m = \ln n + C$ when n = 1 m = mass of N akashon = M_0

Therefore, $C = \ln(M_0)$

To make the value of C more general put $C = K_0 + \ln(M_0)$

K_0 is whole number,

For super heavy akashon K = 0

For electron and particle lighter than electron $K_0 = 2$

For particles heavier than electron $K_0 = 1$

Putting the general value of C in the expression for the value of

$m = (M_0/n)e^{K_0}$

n is huge number $n = 10^x/k$

x and k are small numbers.

Evolution of Physical Laws

n is written in this form because n may not be whole number.

Finally, mass formula is $m=(M_0/n)e^{K_0}$

The value of M_0 is equal to $\sqrt{hc/4\pi G} = 8.63148 \times 10^{21}$ MeV

For electron $K_0 = 2$,

Therefore, putting the value of M_0 and e^2 the mass of electron = $k.M_0 e^2/10^x = (63.77.k/10^x) \times 10^{21}$

For electron mass $k = 8$ and $x = 24$, putting these values calculated mass of electron = 9.097×10^{-31} MeV

The calculated mass of proton = $(kM_0/10^x)e$, because for proton $K_0=1$

Putting the values of constants, the mass of proton $(23.46 \times 10^{21}) \times k/10^x$ S

Put $k = 4$ and $x = 20$ the calculated mass of proton = 1.6734×10^{-27} MeV

Experimental mass of proton = 1.6725×10^{-27} MeV

For particles heavier than electron the particle can be considered to be composed of a number of electrons and a fragment = $kM_0 e/10^x = (23.46.k/10^x) \times 10^{21}$ MeV

Total mass = y(mass of electron) + $(23.46.K10^{21})/10^x$

y is number of electron.

In the brackets experimental mass is given,

For mass of muon $y = 10$, $k = 4$, $x = 21$

Calculated mass of muon = 104 MeV (104)

For Mass of Pion $y = 90$, $k = 4$, $x = 21$

Calculated mass of Pion = 139.7MeV (139.5)

For mass of Kayon y = 16, k = 21, x = 21

Calculated mass of Kayon = 500MeV (500)

For mass of Eita y = 16, k = 20, x = 21

Calculated mass of Eita = 551MeV (550)

For mass of Rho y = 40, k = 32, x = 21

Calculated mass of Rho = 771MeV (770)

For masses heavier than proton the following relation is used,

$yp + (23.46 \cdot k/10^x) \times 10^{21}$

y and p are the numbers and mass of proton

For mass of Taon y = 1, k = 4, x = 20

Calculated mass of Taon = 1776MeV (1777)

For mass of f_2 y = 1, k = 37, x = 21

The calculated mass of f_2 = 1804MeV (1809)

For mass of p_5 y = 2, k = 21, x = 21

Calculated mass = 2368MeV (2359)

For mass of X y = 3, k = 16, x = 22

Calculated mass of X = 2851MeV (2850)

For mass of $D_\pm$ y = 2, k = 4, x = 20

Calculated mass of $D_\pm$ = 1876MeV (1869)

For mass of λ_c^+ y = 2, k = 17, x = 21

Calculated mass of λ_c^+ = 2275MeV (2284)

For mass of λ y = 1, k = 21, x = 21

Calculated mass of λ = 1125MeV (1115)

For mass of W Boson y = 85, k = 21, x = 21,

Calculated mass of W Boson = 80200MeV (80200)

For mass of Z Boson y = 97, k = 8, x = 21

Calculated mass of Z Boson = 91175MeV (91170)

The value k in mass calculation = 2, 4, 8, 16, 32, 17, 20, 21, 37 = $2, 2^2, 2^3, 2^4$, 17, 20, 21, 37…….. 17, 21 and 37 are prime. The value of x=20 or 21only in case of electron it is 24.

The heavy M_N particles may be present in black holes or dark matter. The masses of all the elementary particles are fraction of the mass of M_N Akashon.

The mass of proton can be considered to be composed of infinite concentric Akashons as given below:

Mass of proton m_p = 4.e x 10^{-20}.N.{1 + (1/1!) + (1/2!) + (1/3!) + …….}

Similarly, mass of electron can also be expressed in the above series. Perhaps, this is the cause of stability of electron and proton.

So, many elementary particles are known because the radius of sphere of tranquillity can have many values.

The mass of heaviest elementary particle = $\sqrt{h.c/4\pi G} = 1.539 \times 10^{-8}$ kg

The radius of M_N is named as the smallest measurable length

$R_0 = \sqrt{hG/4\pi c^3} = 1.14 \times 10^{-35}$ m

The formula for the smallest length according to string theory = $\sqrt{hG/c^3}$
= 1.6162×10^{-35} m

The smallest time interval is the time between two successive oscillations of shell situated on the surface of sphere of tranquillity of N elementary particle.

$4\pi R_0.\nu = c$

$4\pi R_0.(1/T_0) = c$

$T_0 = \sqrt{4\pi hGc^5} = 1.9112 \times 10^{-43}$ s

According to String theory, smallest time interval is $\sqrt{hG/c^5}$

The smallest values calculated by me are more accurate and also associate meaning to the smallest length and time interval.

The elementary particles are of two types, fermions and bosons, fermions are composed of Akasha shells and hence offer obstruction, when moving in the opposite directions while bosons are composed of anti-Akasha shells. Anti-Akasha shells are more subtle; therefore, bosons do not offer any obstruction.

Sizes of Electron and Proton:

The radii of spheres of tranquillity of electron and proton are given by

$r_P = h/4\pi m_p c$

$r_e = h/4\pi m_e c$

The effective size of electron or proton depends on the effective presence of region of perturbation which starts outwards from the surface of sphere of tranquillity and is limitless with rapidly tapering off the density which leads to the conclusion that no exact size can be ascribed to electron or proton.

From the mass equations,

$m_p/m_e = r_e/r_p$

The above equation shows that ratio of masses of proton and electron is inverse of the ratio of radii of spheres of tranquillity or roughly sizes. The above result has been experimentally proved.

Why only two types of stable elementary particles are known. As already discussed that region of perturbation of stable particles is composed of hybrid charge mass shells in stretched state. The centers of the stretched shells are shifted in direction opposite to the direction of cosmic force. As the center of the shell touches the surface of the shell, further push or pull is stopped. In case of electron hybrid charged mass centers are present, further inwards growth is stopped at comparatively more distance from the center of the electron hence the radius of sphere of tranquillity is comparatively more in electron, hence, its core is relatively empty of mass shells but is populated by negative charge shells combined with inactive mass shells(without oscillation). In case of proton, region of perturbation is composed of two types of shell upper part is composed of hybrid positive charge mass shells like that of electron while inner part is composed of pure positive mass shells which can be compressed more than that of charged mass shells hence radius of sphere of tranquillity of proton is less than that of electron. Therefore, proton is heavier than electron.

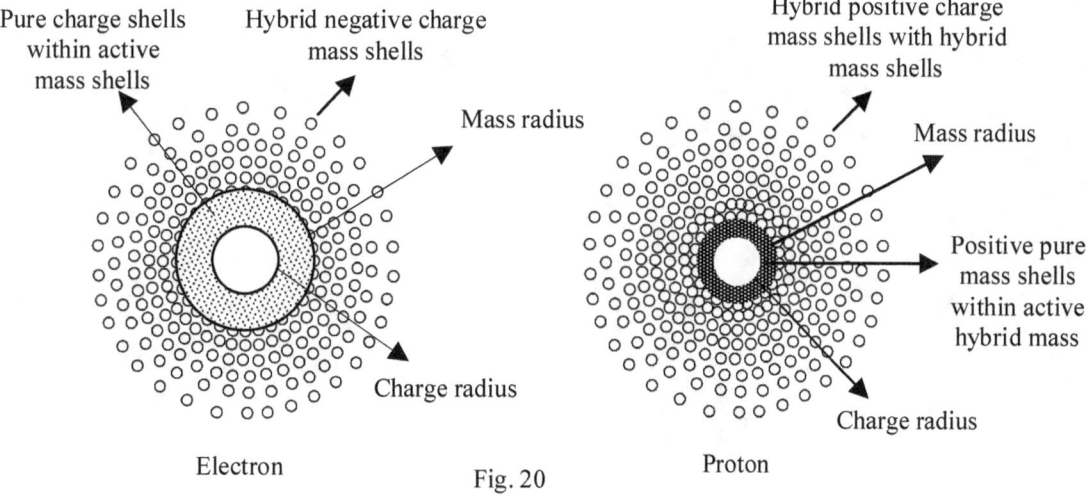

Fig. 20

The structure of proton is different from electron because it is composed of two types of shells but both of these have got relatively empty cores like coconut. There are two radii – charge and mass radii. Charge radii of both electron and proton are expected to be not much different.

QUARKS:

So far it is assumed that the quarks are the building blocks of elementary particles but actually the extremely small shells forming the particles are the ultimate cosmic building material of the particles of the manifested world. The origin of mighty oak or gigantic whale lies in the tiny seeds.

In the elementary particles the majority of shells are confined to region located near the surface of sphere of tranquillity where these shells are held together by very very strong force and therefore have got limited motion. The charge on the elementary particles is also confined to the above region. When elementary particle is struck by powerful projectiles, the group of shells appears as bump. The bump is still part of the elementary particles in the same manner as is the formation of bubbles on the liquid surface. Only very very strong impact of the projectile can separate the bump to give it independent existence. The bumps are named as quarks. The bumps or the quarks are not associated with fixed masses because of lack of independent existence.

The temporary existence of quarks is revealed when the projectile strikes the hadron after following special path. The six lines which are mutually perpendicular in three dimensional space are got by joining the center of sphere of tranquillity of hadron to six symmetrical points fixed on the surface of sphere of tranquillity.

Supposeenergetic conditions are such that six quarks appear on points lying on six vulnerable directions. The charge on each fragment will be one-sixth of the total charge but the six small fragments are highly unstable, therefore, small fragments combine to form bigger fragments. The small fragments prefer to combine in pairs. The two or four fragments can combine but not five because in that case, one unpaired fragment is left. The above observation shows that the charge on the quark can be 1/3 or 2/3 of the charge on electron.

Confinement of Quarks:

The independent existence of quarks require very huge energy to isolate it form the hadron, suppose, one quark is located at a point on one of the six vulnerable directions represented by six lines. As the weak force or disintegration force tries to free the quark from hadron, the mass center of quark shifts toward center of hadron. Therefore, as the quark is pulled out, the confinement force becomes stronger and stronger. The weak force is effective only in separating the two Akashons.

THE FRACTIONAL CHARGE ON THE QUARKS:

In an isolated elementary particle, the charge is confined to the same region as are the mass shells. When subjected to charged projectile, the charge tries to go inside the shell. In the first step, it is distributed over the surfaces of shell and in the second step the charge sneaks into the shells. Suppose r is the radius of each shell and each shell will have two types of charge densities, surface density while it is distributed over the surface of the shells; and bulk density, when the charge is inside the shells. At the surface of the shell,

surface density = r .(bulk density)

r is used to make the dimensions on both sides of the equation same.

It T_s and T_b are the total surface charge and bulk charge, the density relation is changed as given below:

$T_s/4\pi r^2 = r.T_b/ (4/3)\pi r^3$

$T_b = (1/3).T_s$

Surface charge behaves like the charge distributed as in the charge packet, but when the charge is pushed inwards into the shell, its influence becomes one third of the normal charge.

THE STABILITY OF ELEMENTARY PARTICLES:

The electron and proton are stable because these are composite Akashon along with the stretched anti-Akashon shells which impart charge to these particles. In compound elementary particles one lighter and other heavier elementary particles are arranged in such a manner that the lighter elementary particle having large radius of sphere of tranquillity envelopes the heavier elementary particle. Due to oscillations in the region of perturbation the mass centers of the two particles are not concentric.

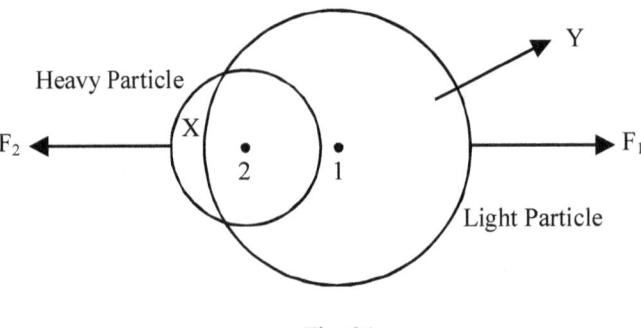

Fig. 21

1 and 2 are the mass centers of light and heavy elementary particles. Due to shifting away of the mass centers the sphere of tranquillity Akashon (1) lies inside the region of perturbation of Akashon (2) and vice-versa. Therefore, oscillating shells of (1) tend to accommodate portion x of the sphere of tranquillity (2) while oscillating shells belonging to (2) tend to be placed inside the region Y of sphere of tranquillity of (2) but it is not the nature of sphere of trqaquillity to accommodate oscillating shells (2) from Y region. Therefore, shells of (2) from Y region are expelled out to region of perturbation towards right arrow and similarly oscillating shells present in space X are expelled out towards left arrow. The expulsion of oscillating

shells leads to disproportionate population of regions of perturbation of (1) and (2). The cosmic force acting on anti-Akasha shells is in outward direction, therefore due to disproportionate distribution of shells particle 1 is pushed from left to right while particle 2 is pushed from right to left, in other words compound elementary particle breaks into two particles, the cause of break-up is weak electrical force. Therefore, here there is again example of unification of weak force and electromagnetic force;in fact weak force is nothing but Coulomb's force. The smaller the distance of separation of mass centers of two elementary particles composing the compound particle the greater is the stability of the compound particle.The weak force is repulsive in nature which helps in separating the fragments of compound particles.

Neutron:

Neutron is compound particle composed of proton and electron. The proton lies inside the sphere of tranquillity of electron because radius of sphere of tranquillity of electron is more than that of proton. The mass and charge centers of proton and electron do not overlap therefore electron is expelled out of neutron in the same manner as explained while discussing the stability of elementary particles. The neutron in the nucleus is stable because as explained later on while discussing the nucleus that neutron is not present as such. During expulsion of electron energy is generated which is carried away in the form of neutrino. In case of hydrogen atom the proton lies inside the sphere of tranquillity of electron which is possible due to large size of sphere of tranquillity of electron. Mass and charge centers of electron overlap and hence hydrogen atom is stable. The hydrogen atom is not under strain hence it is lighter than neutron.

Neutrino:

The shell structure of neutrino has been already discussed. We have seen that charge packet is composed of stretched anti-Akasha shells. The electric field is associated with charge packet. Just like charge packet, the neutrino is composed of stretched shells of Akasha (field effect) but no oscillations (no mas effect). The neutrino is associated with gravitational field. It is counter part of charge packet. Due to absence of oscillations, neutrino shows no interaction and hence has got very large penetrating power. Neutrino can be said to be unbaked Akashon in the cosmic confectionary shop. The anti-neutrino is packet of stretched anti-Akasha shells. Therefore, it is unbaked photon. The rest mass of neutrino is not absolute zero like the photon. The neutrinos are simply carrier of excess energy.

The Dark Matter and Dark Energy:

The dark matter is composed of unbaked Akashon (neutrino) while dark energy is composed of unbaked anti-neutrino. There is nothing mysterious about dark matter or dark energies because these are simply left over of the ingredients used in creation process.

From the above series, it may be concluded that the masses of electron and proton are concentrated in concentric spheres upto infinity with reducing masses. The mass is also present in composite state as in case of true Akashon. Perhaps this special distribution of mass gives permanent stability to proton and electron. The mass formula of super heavy Akashon or unique Akashon can be derived by using dimensions of important constants. Multiplication of the universal constants G, μ_0, ε_0, h and c in terms of dimensions leads to the interesting product from which, mass of the super heavy Akashon or unique Akashon can easily be deduced. The nature is very prolific in supplying building units to generate anything. The vast abundance of pollen during pollination can be cited as an example. It has been seen that the basic building units of mass and energy are shells fashioned out of ether or Akasha. Most of the shells were fixed to form mass packets or photons according to certain rules as discussed earlier but huge left over of the shells were left unused. The leftover shells are used in forming the dark matter, the dark energy and neutrinos.

$h.G.\mu_0.\varepsilon_0 / c = [L^2]$ = Area = constant

The dimension of the product of the constants is that of area. Let us see what is that special area, which is also constant. The special area is equal to the surface of the sphere of tranquility of super heavy N Akashon because it has got constant radius. Let this radius be equal to r_0.

$\mu_0.\varepsilon_0 = 1 / c^2$

Put this value in the product of constants
$h.G.\mu_0.\varepsilon_0 / c = h.G / c^3 = 4\pi r_0^2$
$r_0 = (1/c)\sqrt{(hG / 4\pi c)}$

Evolution of Physical Laws

The mass = $m_0 = (h / 4\pi r_0 c) = \sqrt{(hc / 4\pi G)}$

The above method is very simple.

While getting the mass relation, universal constants G, μ_0, and ε_0 are used. These constants are associated with three fields – gravitational, magnetic and electrostatic. μ_0 and ε_0 are associated with single electromagnetic field. From the above observation it is logical to conclude that with super heavy N Akashon, gravitational and electromagnetic fields are associated or it can also be said that super Akashon unifies gravitational and electromagnetic fields. The unification results through the transitory state of the super N Akashon when it is fragmented into smaller Akshons or photons.

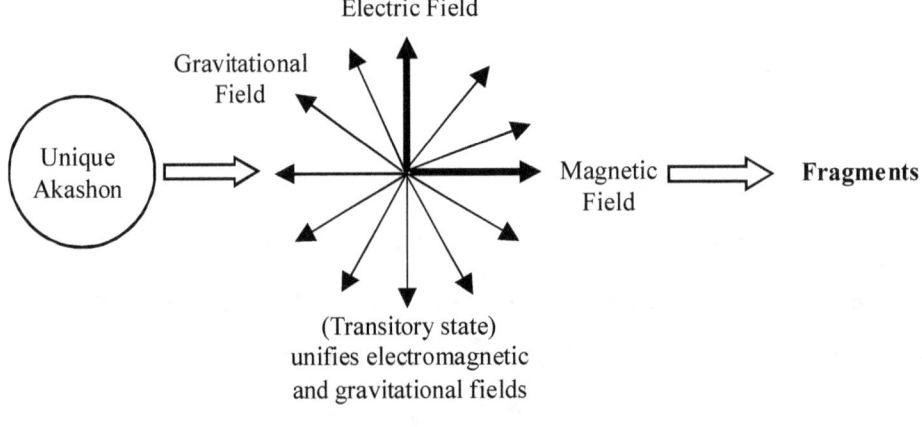

Fig. 22

the radius of sphere of tranquillity of electron $r = h / (4\pi mc)$.

Put the value of constants and mass of electron, $r = 1.9 \times 10^{-13}$m.

Similarly, radius of proton $r = h / (4\pi mc) = 1.05 \times 10^{-16}$m.

From above it seems that radius of proton is less than that of electron. So, it should show more penetrating power than that of electron, but practically it is not so. The reason behind is this that the core of the electron is more vacant. So, obstacles like proton etc. can pass through the electron.

Calculated value of Maximum Force:

The maximum limiting force F_0^* experienced by the unique Akashon

$$F_0^* = \pi h / 16 c t_0^2.$$

When the two unique Akashons during mutual interaction experience limiting force, the surfaces of the spheres of tranquillity of both the Akashons touch each other and mass centers are also simultaneously pushed away to just touch the surfaces of the respective spheres of tranquillity which leads to the fusion of Akashons. After fusion, the mass centers become the part of the region of perturbation, therefore, during the last stage of fusion when the mass centers are moving away under the influence of force for inertial time interval and when the mass center just touch the surface, the oscillating shells present in the surface must be in the mean position which means the inertial time interval for the unique Akashon is equal to half of the time required by the shells on the surface to complete one oscillation.

$t_0 = 1/2\nu_0, \qquad 4\pi r_0 \nu_0 = c$

$4\pi r_0 . (1/2t_0) = c$

$t_0 = 2\pi r_0 / c$

Evolution of Physical Laws

Putting the value of t_0, the limiting force for unique N Akashon is given below

$F_0^* = hc / 64\pi r_0^2$

$r_0 = h / 4\pi m_0.c,$ $\qquad m_0 = \sqrt{hc / (4\pi G)}$

$r_0 = (1/2c)\sqrt{(hG/\pi c)}$

$r_0^2 = hG / \pi c^3$, put the value of r_0^2

$F_0^* = c^4 / 16G = 7.587 \times 10^{42} N.$

Though the value of maximum force is very huge the super heavy Akashon can not move because as already discussed the terminal velocity is equal to zero.

The maximum force is also equal to the gravitational force between two M_N Akashons when $d = 4r_0$ because at the maxcimum force stage mass centers just touch the surface of sphere of tranquillity. Where, r_0 is the radius of M_N Akashon.

$F_0 = G.(M_N^2)/4d^2 = G.(M_N^2)/16r_0^2$

$r_0 = h/4\pi M_N.c$

Putting above values, $F_0 = c^4/16G$

$M_N = \sqrt{(hc/4\pi G)}$

GENERAL GRAVITATIONAL INTERACTION OF TWO AKASHONS:

As seen earlier, it is only in case of mutual interaction of two unique Akashons that both the Akashons simultaneously experience limiting force. As the two Akashons having unequal masses are brought closer, both the Akashons experience maximum force at the same time because maximum force experienced is independent of the initial mass.

Suppose two Akashons with masses M and m are undergoing gravitational interaction and d is the distance between the Akashons at which the interacting Akashons experience maximum force.

At the stage of maximum force, $M = m = M_N = \sqrt{\{hc/(4\pi G)\}}$

$F_0 = c^4/16G$

Gravitational force $= GmM/d^2 = hc/4\pi Gd^2 = F_0 = c^4/16G$

$d = (2/c) . \sqrt{(hG/\pi c)} = 4r_0$

When the above distance which is four times the radius of sphere of tranquillity of the heaviest Akashon is reached, the masses of both the Akashons become equal to the mass of the heavy Akashon M_N. Hence, both the Akashons which are stable before gravitational interaction becomes unstable after the gravitational interaction.

Evolution of Physical Laws

THE VALUE OF k_t:

As already seen the value of inertial time interval t_0 in case of unique Akashon is given by $t_0 = 2\pi r_0 / c$, $t_0^2 = 4\pi^2 r_0^2 / c^2$

$r_0 = (1/2c) \sqrt{(hG/\pi c)}$

Put the value of r_0^2, as already deduced

$t_0^2 = (4\pi^2 / c^2).(hG / 4\pi c^3) = \pi hG / c^5$

$t_0 = \sqrt{(\pi hG / C^5)}$

$k_t = 4\pi r_0 t_0$

Put value of r_0 and t_0

$k_t = 4\pi \sqrt{(hG / 4\pi c^3)} . \sqrt{(\pi hG / c^5)} = 2\pi hG / c^4 = 1.715 \times 10^{-79}$ m.s.

The very very small value of k_t indicates that inertial time $k_t/4\pi r$ for the given Akashon will be significant only if r is very very small which means only very heavy Akashons have inertial time interval which cannot be neglected otherwise for lighter Akashons it can be easily neglected.

GUIDELINES TO OBSERVE LEVITATION OR TO OVERCOME FORCE OF GRAVITY:

It is already seen that cosmic force squeeze inward shells of the Akashon equally in case of isolated Akashon. The cosmic force can be unbalanced if the region of perturbation is bombarded with high energy projectiles or high energy electromagnetic radiation (photons). On bombardment the regular arrangement of shells will be jumbled resulting in unbalancing of the cosmic force because that side of the bombarded region of perturbation will be squeezed inward with less force than the other region. Therefore, the unbalanced force pushes the Akashon in the direction from un-bombarded region to the bombarded region.

SPOOKY ACTION AT DISTANCE IN PLANTS (THE PROOF OF EXISTENCE OF ETHER):

In defense of presence of Akasha or ether and the conclusion that region of perturbation of Akashon (mass packet or photon) is limitless.

As has been discussed earlier that mass packet and energy packets are composed of ether or Akasha. The bonding surfaces of the objects are only apparent because each object has got limitless dimensions. The density of ether shells falls off rapidly with increase of distance from the apparent surface, that is why the object appears to have non-existent boundary.

Goethe said, "One perceives the fundamental essence of life in the living, not the inanimate, in that which is changing, not what is finished".

The experiments on plants and spooky action at a distance clearly prove the existence of ether or Akasha. The following are simple experiments and observations supporting the existence of ether and infinite dimensions of objects.

Experiments on plants (The experiment on buds was performed for the first time):

1. **Movement of flower buds:** The experiment on flowers' buds shows the spooky action at distance. The experiment was performed by selecting buds of Marigold and other plants like that of Asteraceae family having naked and terminal buds with long stalks. The flower buds show the effect of actiona at distance. The experiment does not

need any sophisticated instrument but the results **clearly prove the existence of ether beyond any doubt**. It is better if the experiment is performed in the morning and the plant is well watered. The buds selected should be young and healthy.

The Experiment:

The pencil like object or rods with small diameter attached to vertical sticks were fixed horizontally in such a manner that the one end of the pencil did not actually touch the bud, but was kept at 2-3mm away from the bud (see the diagram).

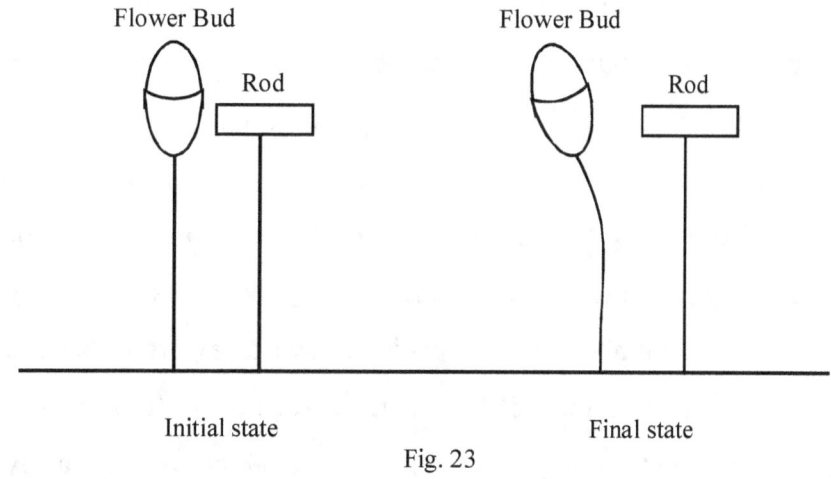

Fig. 23

The experiment was performed simultaneously on the buds of the same plant. The rod ends were arranged all around the pot containing the plant so that the ends pointed towards the buds in multiple directions (360^0) without preferring single direction. After 2-3 hours interval, it was observed that buds started moving away from the pencil ends, time interval varies with the temperature and state of hydration of the roots. More than 80% buds experienced the motion away from the rod end. Other buds might not be healthy. The nature of the material of the rod wood, plastic, rubber, metal, and magnetdid not influence the result of the experiment. The results were the same when the experiment was performed in the dark room.

Explanation for the observed effect:

The flower buds open up to form flowers which further change to fruit which culminate in the formation of seeds and with that the goal of the plant growth is achieved, therefore the bud is the most prescious gift of nature. The bud is expected to be very sensitive to the presence of intruder in its vicinity. The motion of buds is not due to nastic movement of the buds because nastic movement is non-directional and random in nature and it is also not possible that all the buds should experience nastic moment in the same direction. The movement of the bud is also not due to negative thigmotropic motion because rod does not actually touch the buds but are kept at a little distance away from the bud. The number of buds which move away from the rod ends always much more than 50% which shows that shifting away of the buds is not due to nastic movement. The movement of buds away from the pencil ends is not negative thigmotropic motion because the pencil ends did not actually touch the buds. It is an example of **spooky action at a distance**. The length of pencil actually does not end at the apparent end it is extended beyond the apparent boundary. There is effective existence of the rod and bud between the gap which separate the bud from the apparent pencil end the gap is not empty. It is filled with the extra presence of ether shells associated with the pencil and bud. These shells initiate the reactions in the bud to bend it away from the pencil end and thus creating more space for it to blossom into a flower because much space is needed when the bud changes into flower. The experiment proves the existence of ether and limitless dimensions of all the objects. The same results are retained if the buds are wrapped in metal foils which eliminate radiation of any waves.

The above simple experiment clearly shows the limitless dimensions of all the objects and hence presence of Akasha or ether.

2. **Experiment on tendrils ofcreepers:**

If the growth of creeper growing at the base of wall is observed, it is seen that much more than 50% of tendrils are bending towards the wall and growth of the creeper is towards the wall even when the creeper or tendrils have not yet actually touched the wall.

The observation can not be explained by the nastic moment of the tendrils because nastic movement is without any direction. The presence of external objects introduces the property of direction to the nastic movement. The presence of external object is felt by the temdrils through the presence of extra Akasha or ether shells associated with the objects or tendrils. Similarly, the thigmotropic property of *mimosapudica* can be explained with the help of extra shells of Akasha of ether. The touching of the plant introduces extra shells to the *mimosapudica* plant leaeves and the presence of these extra Akasha shells is responsible for the thigmotropic movement experienced by the plant.

3. **The direction of growth of branches of the trees or plants:** If a tree or plant with the branches is observed it is seen that nearly all the branches which are attached to lower part of the stem are bow shaped or curved upwards, while branches higher up are straight. One is first tempted to explain the above observation by using the concept of geotropism but it cannot be so because force of gravity at two points vertically separated by small distance is very nearly the same because it is inversely proportional to the square distance from the center of the earth. The interaction of two regions of perturbation of Akahsons present in the surface of the earth and branches of the tree or plant can be used as explanation. The interaction depends on the shell density of Akasha or ether of the region of perturbation which is equal to $h/16\pi^2 cr^4$, r being the distance between the earth surface and branch. The above relation proves that interaction unlike gravitational interactionfalls rapidly with increase in distance. The branches attached to the lower part of stem sense the near presence of earth surface more than the branches attached to higher points. The natural tendency of the plant or tree is grow upwards therefore, the branches at lower part try to evade the surface of the earth by acquiring curved shape. In case of tendrils the natural tendency is to attach to a support, therefore, sensing the nearby presence of support through interaction of regions of

perturbation of tendrils and the nails, the tendrils change their path of growth in the direction of support.

4. **The case of entangled photons:** that entangled photons can influence each other when separated by even by large distance is due to the presence of limitless region of perturbation. The same type of entanglement is also possible in pair of electrons.

5. **Simple experiment to detect the direct presence of regions of perturbation in photon and electron:** take simple glass tube. The ends of the tube are coated with material to detect the presence of photons and electrons on striking the opposite ends. Inside the tube on the opposite ends of the tube sources of emission of beams of strong photon (laser) and electrons are fitted. The two beams travel parallel to each other in opposite directions. The thickness of the beams is not so less that marked diffraction fringes are produced while passing through small holes placed before the respective sources. The tube is evacuated to remove air. To start with first only one beam is turned on and its position on the screen is noted. After that both the beams are turned on simultaneously to observe shift in the spots marked by the beam when only one beam was turned on and afterwards when both beams are moving in the opposite direction simultaneously. The shift in the beam will be produced due to lateral interaction of the regions of perturbation, photon, and electron. There should be no head-on collision of the particles of the respective beams.

The above experiments and observations leaeve no doubt about the existence of ether.

THE CHARGE AND ELECTRIC FIELD:

The charge packet is formed by the clustering of the outwards stretched or displaced shells of the positive or negative anti-Akasha around the central sphere of tranquillity. In case of charge packets anti-ether or anti-Akasha shells are grouped around sphere of tranquillity of charge packet. There are two types of anti-Akasha +ve and –ve therefore there are two types of charge packets +ve and –ve. Like etheron or Akashon the shells are arranged in the form of springs which are stretched outwards with cosmic force.

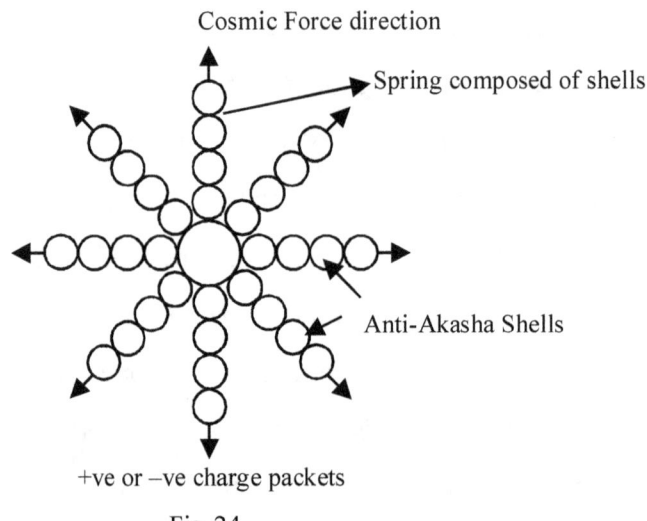

Fig. 24

As in case of Akashon the absolutely isolated charge packet does not experience any force. The mutual interaction of two charge packets unbalances the primary cosmic force, which is manifested as the secondary force called electric field. There are two types of charge packets therefore there will be two types of mutual interaction, between the same type of charge packets and opposite types.

1. **Mutual interaction of same types of charges:** Mutual interaction between same types of charges is of the same form as that of Akashons.

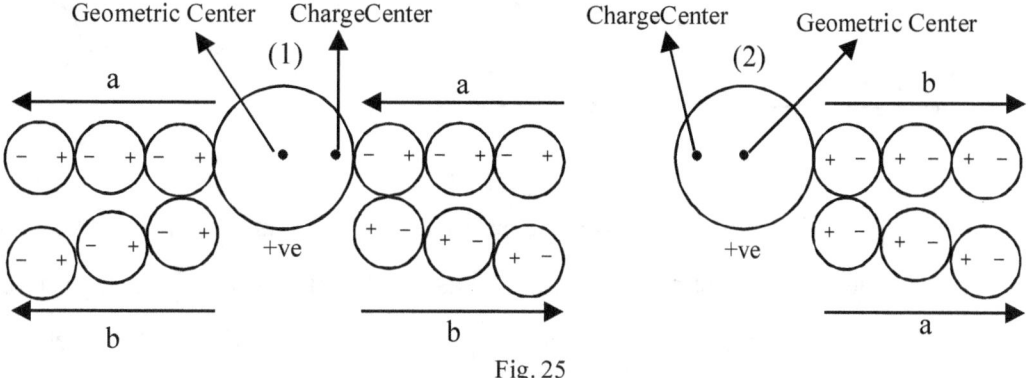

Fig. 25

As is seen from the diagram, due to mutual interaction, the resultant force on the left side of charge (1) is reinforced while on the right hand side it is subdued. In case of charge (2), the case is reversed. Overall result is this that charge (1) and charge (2) are pulled outwards or we can say that similar charges repel each other. Due to unbalancing of the charges, charge centres of both the charge packets also shift away from the geometric centre which undergoes very little change in its position.

2. **Interaction of opposite charges:** here two types of charge packets are to be considered. In case of –ve charge packet, the shells of –ve anti-Akasha are present while anti-Akasha shells of the +ve type are presentin +ve charge packet. The mutual interaction is shown by the following schematic device.

Evolution of Physical Laws

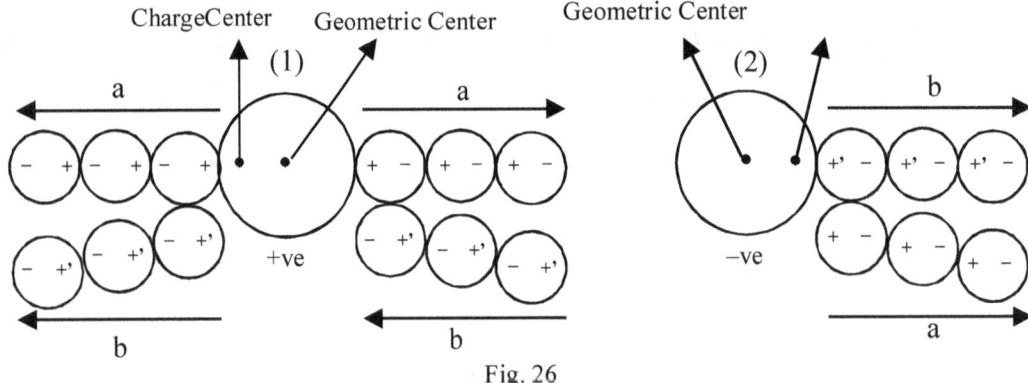

Fig. 26

In the diagram above the excess of anti-Akashais shown by +ve sign while –ve sign shows less anti-Akasha. When +ve and –ve regions of shells overlap, disproportionate distribution of anti-Akasha becomes less because +ve region shows excess presence of +ve anti-Akasha and (+') region shows excess presence of–ve anti-Akasha(+) and (+')cancel each other while(+') and (-)reinforce. As a result of above observation, the force on the left hand side of +1 charge becomes less while on the right hand side, it is reinforced because here there is no overlapping of +ve and (+') regions. The force acting outwards on –2 charge on the right hand side becomes less. The net result is that, the force acting on +ve charge is from left to right while on –ve charge, it is from right to left or it can be said that opposite charges attract each other.Charge centres of both the charge packets are shifted away from the geometric centre. There may be other modes of mutual interaction but the net result of mutual interaction is the shifting away of the charge centre away from the geometric centre.

3. As in case of gravitational interaction due to the expulsion of charge shells present in the spheres of tranquillity of interacting charge packets

the force is also exerted due to disproportionate distribution of expelled shells in the region of perturbation.

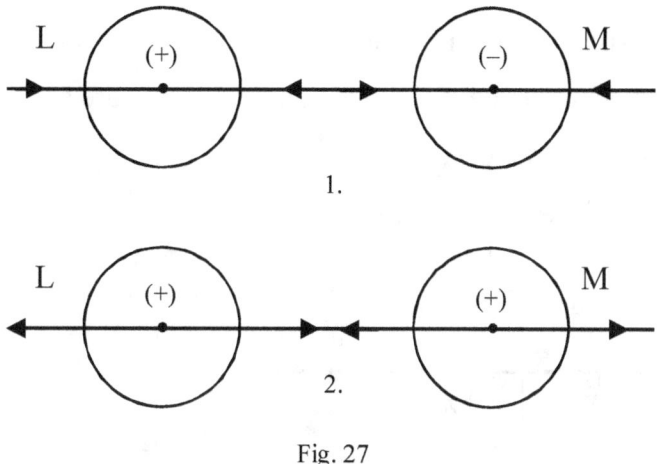

Fig. 27

In diagram 1, L and M are rarer regions due to expulsion of shells of opposite sign hence, attractive force is there. While in diagram 2, L and M are denser regions due to expulsion of shells of same size hence, repulsive force is there.

The electric field is not manifested instantaneously because the readjustment of the shells during mutual interaction requires some time. The above discussion is the simplified explanation of the mutual interaction of the charge packets, whatever may be the explanation, one thing is clear that the primary cosmic force is already there to be manifested as the secondary force called electrostatic field during mutual interaction of charge packets.

Evolution of Physical Laws

DERIVATION OF COULOMB'S LAW OF FORCE BETWEEN TWO CHARGES:

Let us next discuss the magnitude of the force. Suppose, L_A and L_B are the radii of the spheres of tranquillity of the charge packets A and B respectively. Two oppositely charged charge packets are selected to discuss the case of attractive force.

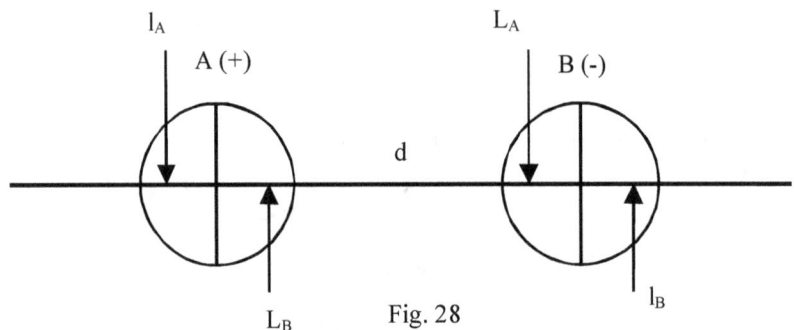

Fig. 28

The force on the each charge packet is given by the usual force relation.
$F_A = F_{0B} \, l_A / L_A$, $F_B = F_{0A} \, l_B / L_B$

The force F_A and F_B are manifested as a consequence of the mutual interaction.

The procedure adopted to calculate the electrostatic force is the carbon copy of the treatment used to calculate the gravitational force; the only difference is the change in the magnitude of the constants involved. The force acting on the given charge packet depends on the three factors, the force factor, the distance of separation and the radius of the sphere of tranquillity. As in case of Akashon, the principle of inverse variation is applicable to determine the values of the force factor and the displacement factor.

$F_{0A} = F_{0B} = K_E'/4\pi d$

$F_A = (K_E'/4\pi d).(l_A/L_A)$

$F_B = (K_E'/4\pi d).(l_B/L_B)$

As in case of interaction of Akashons

$l_A/L_A = l_B/L_B,$ therefore $F_A = F_B$

$l_A = k'q_B,$ $l_B = k'q_A$

q_A and q_B are the magnitude of the charge on the charge packets. As in case of Akashons the displacement factor k' can be determined by the application of the principle of inverse variation

$4\pi dk' = k_0'$ therefore,

$l_A = k_0'q_B/4\pi d$; $l_B = k_0'q_A/4\pi r$

$4\pi L_A q_A = $ constant $= k_q$ and,

The magnitude of hybrid charge is calculated by applying the principle of inverse variation

$4\pi L_B q_B = k_q$ therefore,

$F = F_A = F_B = (K_E'.k_0'/4\pi k_q). q_A. q_B/d^2 = q_A. q_B/4\pi \epsilon_0.d^2$

$\epsilon_0 = k_q / (K_E'.k_0')$

Using the dimensions of constants it can be shown that if ϵ changes then to the following relation inverse variation relation is applicable

$4\pi r\epsilon = $ constant.

Here, r is the radius of sphere of tranquility of charge packet and ϵ is the value on the surface of sphere of tranquility. This important reault will be used further.

The Coulomb's Law of Force has been derived in very simple manner. The particle exchange (virtual photon) concept of field has failed to derive the above force relation.

When the two charge packets interact, the charge or force centre is displaced away from the geometric centre. The temporary decrease in the radius of sphere of tranquillity brings about increase in the electrical energy stored in the region of perturbation. The extra energy stored is carried away by the Prakashon (photon), therefore, there is no permanent increase in the charge. No exchange of virtual photons or virtual gravitons is required for the manifestation of electric or gravitational field. The range of electric and gravitational field is boundless in both the cases. When two charge packets interact, the force manifested as electrostatic force is actually the unbalanced cosmic force in the same manner as gravitational force is manifested. Therefore, gravitational field and electrostatic field are unified because the cause of origin of both the force is common, cosmic force.

While discussing the mutual interaction of the Akashons (mass packets) or charge particles, three factors were considered which bring about change in the state of shells present in the region of perturbation and out of these three factors, it may be that only one factor plays dominating role in shifting away the mass centre or the charge centre. The expulsion of the shell out of the sphere of tranquillity of the interacting Akashons or charge packets may play dominating role in unbalancing equal distribution of shells around the sphere of tranquillity and hence lead to shifting away of the mass centre of the mass packet or the charge centre which causes the manifestation of gravitational or electrostatic field. As already seen this factor alone is sufficient to explain the attractive force between the Akashons and repulsive and attractive force between charge packets associated with the similar and opposite types of charges.

Like gravitational constant permittivity is also natural property and its value is given by the principle of inverse variation $4\pi r \epsilon = k_\epsilon$.

ϵ is the permittivity at a point situated on the surface of sphere of tranquillity at distance r from the charge center. At large distance the rate of change of ϵ with change in distance is nearly zero. So it can be taken as constant, but at very small distance, change in value of ϵ with distance change can not be neglected.

$\epsilon_0 = k_q / k'_E k'_q k_q = (4\pi \text{charge.length})$,

$k'_E = (4\pi \text{energy})$

The constants have been written in terms of dimensions. Putting the values of constants $\epsilon_0 = \left(\dfrac{(\text{charge})^2}{4\pi \text{energylength}}\right) 4\pi(\text{energy})(\text{length}) = hc$ is the maximum energy packed in special energy packet or prakashon. The relation between charge and energy proves that enegy packets or photons are related to charge. It shall be proved later on that photon is simply doublet of pure charge there is nothing enigmatic about it. Putting the value $4\pi(\text{energy})(\text{length}) = hc$, the value of maximum charge $= Q_0 = \sqrt{hc}\,\epsilon_0$, the above is the value of maximum hybrid charge. Calculated value of maximum hybrid charge $=13.267 \times 10^{-19}$ C. The value of permittivity on the surface of sphere of sphere of tranquillity of charge packet is constant and that value is the conventional value. It will be seen later on that the expression for pure charge is of the same form the approximate value of charge on electron $=\sqrt{hc\,\epsilon_0/22\pi}=1.5957 \times 10^{-19}$ c. From the values of maxim um mass and charge expressions it can be easily proved $G_0 \times \epsilon_0 = 4\pi Q_0^2/M_0^2$ It can be said that constant of gravity associated with gravitational field and permittivity associated with electrostatic field are unified through maximum charge packet and maximum mass packet N akashon.

CALCULATION OF ELEMENTARY CHARGE AND TYPES OF CHARGE:

The classical definition of unit charge is only arbitrary. Let us next discuss what is the most natural way of defining unit charge if the new concept of charge as discussed in the present text is followed. The charge can exist either as pure charge packet in which only stretched anti-Akasha shells are present. The second type of charge is blended charge packet in which stretched Akasha cells (mass) are associated with stretched anti-Akasha shells (charge) to form hybrid of mass charge shells. Thus it is seen that in blended charge packet hybrid charge shells are stretrched while in pure charge packet only pure charge shells are stretched. There is a limit to the magnitude of hybrid or blended charge while pure charge magnitude is variable. It will be seen later on that pure charge is present in photon. Phtons contain two pure charge packets carrying equal and opposite charges with oscillation of the charge shells. In case of hybrid charge packets the radius of sphere of tranquillity is fixed ($4\pi rq = k_q$) hence charge is fixed but in case of pure charge packet, radius is variable hence magnitude of charge is variable.

The extent of stretching of pure charge shells is much greater while in case of blended charge packets, there is limit to the extent upto which blended charge shell can be stretched, hence the magnitude of the blended charge packet is fixed. The concept of permittivity also differentiate the behavior of two types of charges regarding the magnitude. According to the classical concept, the value of permittivity is same at each point. The value of vacuum permitivity is used for calculation purpose. But the permittivity being natural property, the value of permittivity at point situated on the surface of sphere of tranquillity of the charge packet is $4\pi r \in = k_\epsilon$. If the permittivities at two points situated on the sphere of tranquillity of radii r_1 and r_2 from the centre of the sphere of tranquillity is to be compared, the principle of inverse variation is applicable. If $\in_1$ and $\in_2$ are the magnitudes of permittivities, the relation is as follows:

$r_1 \in_1 = r_2 \in_2$

The above equation shows that permittivity is natural property but the usual value assigned to vacuum permittivity is only arbitrary value. The radius of sphere of tranquility of hybrid charge is fixed, hence the value of permittivity at a point on the sphere of tranquility is fixed, but in case of pure charge packet r is variable hence both charge and permittivity are variable.

Let us now discuss the force between two pure charge packets when the distance between the charge packets is minimum. The force experienced by the charge packet is maximum. The Coulomb's force relation is valid even up to small distance because the interacting pure charge packets are composed of pure charges hence no other interaction except the charge interaction is there. The maximum force relation is taken to be the same as in case of gravitational interaction of two pure Aksahon.

Maximum limiting force = $\pi.h / 16.c.t^2$
The minimum distance between charge packets = $4r$

r is the radius of sphere of tranquillity of pure charge packet

ϵ_p is the value of permittivity at the surface of sphere of tranquillity

Electrostatic force = $(1 / 4\pi\epsilon_p).(q^2 / 16r^2)$

Here permittivity ϵ_p changes with r because pure charge is being discussed. Equating the two forces,

$(1 / 4\pi\epsilon_p^0).(q_0^2 / 16r^2) = \pi.h / 16.c.t^2$

put the value of $t = 2\pi r / c$

$(1 / 4\pi\epsilon_p^0).(q_0^2 / 16r^2) = (\pi h / 16c).(c^2 / 4\pi^2 r^2)$

$q_0 = \sqrt{hc\epsilon_p^0}$

ϵ_p^0 is permittivity at the surface of the sphere of tranquillity of pure charge packet having radius equal to r. The above equation shows that the value of charge q associated with charge packet goes on changing as the value of permittivity ϵ_p^0 on the surface of sphere of tranquillity goes on changing.

q_0 is the maximum value of pure charge because it is calculated when the force is maximum or its value depends on ϵ_p^0 which is maximum value.

Maximum charge $q_0 = \sqrt{hc\epsilon_p^0}$

The above charge relation is also applicable to calculate pure charge associated with any other pure charge packet,

Evolution of Physical Laws

$$q = \sqrt{h.c\epsilon_p}$$

ϵ is permittivity on the surface of sphere of tranquillity of pure charge packet having radius r. the principle of inverse variation is applicable to permittivity.

$$4\pi r\epsilon = k_\epsilon$$

It will be shown later on that photon or electromagneton is composed of two oscillating pure charge packets carrying opposite charges. With the change in r, therefore, magnitude of pure charge can vary. The pure charge is associated with photon.

Unlike the blended charge which has got one value the pure charge value can vary. The natural unit of charge and permittivity can be defined as follows in case of pure charge packet:

$$q_0 = \sqrt{h.c\epsilon_p^0}$$

Suppose maximum pure charge q_0 is taken equal to 1, the unit charge fixed is natural unit unlike the arbitrary unit of charge called Coulomb.

Put q = 1 in the above equation

$$1 = \sqrt{h.c\epsilon_p^0}$$

$$\epsilon_p^0 = 1.986 \times 10^{25} C^2 N^{-1} m^2$$

Therefore, under above definition pure permittivity is very huge. The value of permittivity can be made less if natural unit of charge is defined by taking q equal to very small number instead of one. The conventional unit of charge is based on force property of charge and arbitrarily fixing the magnitude of force and distance between the conducting wires.

It will be seen that photon is composed of electromagneton. What is electromagneton will be also discussed soon. Doublet of charge packet is present in electromagneton. Each part of the doublet is associated with pure charge equal to $\sqrt{(hc\epsilon_p)}$.

The pure charge present on each electromagneton of photon will be subsequently taken equal to $\sqrt{h.c\epsilon_p}$.

ϵ_p is the value of permittivity at the surface of the sphere of tranquillity of electromagneton.

The natural permittivity ϵ follows the principle of inverse variation.

The pure charge packet is unstable unlike the hybrid charge. Even energy packet carrying one type of oscillating pure charge is unstable. Stability is required when two pure charge packets of opposite types combine to form neutral dublet of pure charge packets. The neutral doublet of oscillating shells of pure charge packets form the photon as will be discussed later on.

METAMORPHOSIS AND CHARGE PACKET:

As already seen in blended charge packet, the shells are stretched outwards and their innermost shells are in state of maximum limiting displacements that is why unlike the Akashons or mass packets, the charge packet with only one unit of charge is known. When the blended charge packet is subjected to acceleration or motion, the shells with more than limiting displacement are created on the periphery of the sphere of tranquillity but that is not allowed, therefore, the newly created hybrid charge shells are metamorphosed into photon shells. In other words there is emission of electromagnetic radiation when the charge packet is accelerated.

SELF ELECTRICAL POTENTIAL ENERGY OF CHARGED AKASHON AND ELECTROMAGNETON:

When neutral Akashon is given positive or negative charge, during the process of charging, self electrical potential energy is pumped in the Akashon. Suppose neutral Akashon composed of positive Akasha mass shells is to be charged. Each mass shell is blended withpositive or negative charge shells to form hybrid mass or hybrid charge shells which are more stable than neutral mass shell. As more and more negative charge shells are introduced hybrid charge on the initially neutral mass packet goes on increasing, hence work has to be done to introduce more negative charge shells. The work done is called electrical self potential energy.The self electrical potential energy pumped in during charging process is stored in the hybrid charge as packet of energy called **electromagneton**.The electromagneton composed of hybrid charge is called permanent electromagneton it is stable and carries only one type of hybrid charge, hence named as singlet state of electromagneton. All the particles like electrons or protons carrying hybrid charge are associated with singlet state of stable electromagneton.The electromagneton like the mattenergon is not virtual but real. It has got sphere of tranquility and region of perturbation but co-exist only with the charge particle. Electromagneton of pure charge has got independent existence. Photon is actually doublet (±) of pure charge electromagneton.The self electrical potential energy when associated with pure charge packet forms unstable singlet state of electromagneton. The singlet state of unstable electromagneton of pure charge at once combines with singlet state of another electromagneton carrying opposite charge to form neutral doublet state which will be seen afterwards is actually photon.

The electromagneton is neutral energy packet. It has got sphere of tranquillity as the sphere of tranquillity of associated charge packet or that of the Akashon and is composed of oscillating charge shells. All the charged Akashons are enveloped by electromagnetons. The electromagneton can share extra energy during interaction. The energy associated with electromagneton is transit form of electromagnetic energy or it is precursor of photonic energy. Unlike the virtual particles the electromagneton is real in nature.

Calculation of Self Electrical Potential Energy:
The calculation of above energy is possible because mass packet and charge packet have got spheres of tranquillity with finite radii. The mass packet or Akashon is not point particle as is the usual convention. The point dimensions of the particles pose the problem of singularity or infinite value which is overcome by use of artificial concept of renormalization, but according to the natural size of the Akashons there is no need to use the artificial concept of renormalization.

Case1: here energy is calculated so that at each point in the region of perturbation while the charge is being pumped in there is no change in the permittivity at a point located on the sphere of tranquillity. Suppose 'e' is the magnitude of charge on the charge packet and r is the radius of the sphere of tranquility at any instant.

$4\pi r e = k_e$

If the charge is supposed to be concentrated at the centre, the electrical potential energy at the surface of the sphere of tranquillity is equal to $e/4\pi \in_0 r$.

The increase in the charge with decrease in the radius of the sphere of tranquillity is given by the following equation
$$de = -(k_e/4\pi r^2).dr$$

The increase in electrical potential energy as a result of change in charge is given by the following equation

$$dE = (1/4\pi \in_0).(e\,de/r)$$

Put the values of e and de
$$dE = -(k_e^2/64\pi^3 \in_0).dr/r^4$$

$$E = \int_0^E dE = -(k_e^2/64\pi^3 \in_0) \int_\infty^r dr/r^4 = (k_e^2/192\pi^3 \in_0)/r^3$$

Put the value of $k_e^2 = 16\pi^2 e^2$ in the above equation
$$E = (e^2/12\pi \in_0).(1/r)$$

The above relation gives the total electrical potential energy stored in the region around the sphere of tranquillity of the charge packet with radius r. The energy packet in which self electrical potential energy is stored is called electromagneton. The above result will not be used because it is wrong.

Evolution of Physical Laws

Case 2: Actually there is change in $\in$ on the surface of sphere of tranquility as radius changes from infinity to final r as charge packet is being formed. In this method, for determining the electrical potential energy pumped into the region of perturbation, the change in the permittivity is taken into account. The permittivity at the points located on the surface of the sphere of tranquillity depends on r. Suppose, instantaneous permittivity at a point situated at a point on the surface of sphere of tranquillity of radius r is to be determined, its value can be known by applying the principle of inverse variation. Instantaneous permittivity inside the sphere of tranquillity is same as at the surface. Suppose, $\in$ is the permittivity at this point and $\in_0$ is the permittivity at the surface of sphere of tranquillity of radius r_0 after the completion of the formation of region of perturbation.

$\in$ and $\in_0$ are the permittivity at a point on the surface of sphere of tranquillity of radii r and r_0. Apply the principle of inverse variation,

$4\pi r \in = 4\pi r_0 \in_0$

$\in = r_0 \in_0 / r$

Suppose r is the radius of sphere of trqnauillity of charge packet at any instant when the region of perturbation is being built up,

Potential Energy $E = (1/4\pi\in).(e/r)$

Suppose extra charge 'de' is added up to the region of perturbation during its formation. If 'e' charge at any instant is supposed to be concentrated at the centre of the charge packet, increase in potential energy

$dE = (1/4\pi\in).(e.de/r)$

The magnitude of charge e at any instant is given by $4\pi er = k_e$

$de = -k_e / 4\pi r^2$

Putting the values of e and de

$dE = -(k_e^2 / 64\pi^3 \in_0 r_0).(dr/r^3)$

$k_e^2 = 16\pi^2 r_0^2 e_0^2$

e_0 is the value of full charge,

put the value of k_e^2

$$E = \int_0^E dE = -(e_0^2 . r_0 / 4\pi\in_0) \int_\infty^{r_0} (dr/r^3)$$

$E = e_0^2 / 8\pi\in_0 r_0$

r_0 is the radius of sphere of tranquility of final hybrid charge packet.

$4\pi r_0 E = e_0^2 / 2\in_0 r_0 =$ constant

The above equation shows that the principle of inverse variation is applicable to find out the electrical potential energy stored in the region of perturbation. The electrical potential energy is stored as electromagnergy in the electromagneton, which has got its sphere of tranquillity equal to the charge packet. The second method for calculating the energy associated with the electromagneton is more accurate than the first case because in the second case change in permittivity is accounted for, therefore the second value of electromagneton energy will be used.

Evolution of Physical Laws

THE ACCELERATION OF THE CHARGED AKASHON:

Why an accelerated charge emits radiation has not been understood clearly so far. The various proposed concepts are confusing. The reason for emission of radiation proposed in this text is very easy to understand. When the charged Akashon (electron) is accelerated, there is decrease in the radii of both the Akashon and charge packet. Suppose r and r_0 are the radii of the moving charged Akashon and at rest, the change in energy of the electromagneton is given by the following equation. The case 1 is incorrect, therefore, no change in energy will be considered for that case.

Case 2: The energy of the electromagneton when electron is at rest

$$E_0 = (e^2/8\pi \in_0)/r_0$$

When electron is set into motion $r_0 \rightarrow r$, $r < r_0$

$$E = (e^2/8\pi \in)/r$$
$$\Delta E = E - E_0 = 1/8\pi . \{(e^2/\in r) - (e_0^2/\in_0 r_0)\}$$

The above is the increase in the value of electromagneton energy as result of motion.

There is temporary change in values of e and $\in$ as r_0 changes to r on account of motion. Suppose v is the velocity of the Akashon. Tiny change in charge carried away by photon.

$r = r_0 . \sqrt{\{1-(v^2/c^2)\}}$

The changed values are given by the application of principle of inverse variation.

$r e = r_0 e_0$

As $r_0 \to r$, $e_0 \to e$ thus, there is small increase in hybrid charge but hybrid charge has got fixed value. The excess hybrid charge changes to pure charge which is associated with temporary region of perturbation as doublet ($\pm$) and is carried away by photon. Therefore there is no increase in the charge of accelerated hybrid charge particle.

$r\epsilon = r_0 \epsilon_0$
$e^2 = r_0^2 . e_0^2 / r^2$
$e^2/\epsilon r = r_0^2 e_0^2 / r^2 r_0 \epsilon_0 = r_0 e_0^2 / r^2 \epsilon_0$

$\Delta E = (1/8\pi) . \{(e^2/\epsilon r) - (e_0^2/\epsilon_0 r)\} = (1/8\pi) . \{(r_0 e_0^2 / r^2 \epsilon_0) - (e_0^2/\epsilon_0 r_0)\}$
$= (e_0^2 / 8\pi \epsilon_0 r_0) \{(r_0^2/r^2) - 1\}$

$r = r_0 . \sqrt{\{1-(v^2/c^2)\}}$

Put the changed values,
$\Delta E = e^2/8\pi\epsilon_0 r_0 . [\{1-(v^2/c^2)\}^{-1} - 1] = e^2/8\pi\epsilon_0 r_0 . (1 + v^2/c^2 + v^4/c^4 + \ldots - 1)$
When, $v \ll c$, $\Delta E = (e^2/8\pi\epsilon_0 r_0) . (v^2/c^2)$

It has already been proved that square of velocity is quantized $v^2 = 2Nc^2/n_0$
After putting the above value,

$\Delta E = (e^2/8\pi\epsilon_0 r_0) 2N/n_0$
$N = 0, 1, 2, 3, \ldots\ldots$

The relation shows that energy of temporary electromagneton or electromagnetic radiation is quantized contrary to Larmor relation which wrongly specifies that there is continuous emission of radiation.

The above is the additional energy of unstable singlet electromagneton which at once changes to doublet electromagneton which is actually photonradiation. Electromagneton associated with temporary region of perturbation formed as a result of acceleration is unstable. It is this additional electromagneton energy which is changed to electromagnetic radiation. Suppose electron is accelerated from rest with acceleration a for time t,

$v^2 = a^2.t^2$
Therefore, $\Delta E = (e^2/8\pi\epsilon_0 r_0)(2N/n_0)$

Time interval for emission of photon and invariance of light speed:
When electron is accelerated the sphere of tranquillity of temporary electromagneton subsequently changes to photon starts forming. Though electron is moving the centre temporary elctromagneton remains stationary because it is not part of electron. When centre of temporary

electromagneton just touches the surface of sphere of tranquillity of moving electron the temporary electromagneton is metamorphosed to photon which starts moving with constant terminal velocity c w.r.t to point of origin of centre of photon and thus velocity of photon or light is independent of electon or source velocity. Suppose r is the radius of sphere of tranquillity of temporaty electromagneton or photon. The centre of moving electron will touch the surface of tranquillity of photon in the manifestation stage after travelling distance r with acceleration a during time interval t,

$r = 1/2(at^2)$

if λ is wave length of radiation $\lambda = 4\pi r$

$h\nu = c$

using these relation it can be seen that $t^2 = c/2\pi\nu a$

t is the time interval needed for the formation of photon it has been seen that half of the electromagneton energy is changed to photon hence,

$\Delta E = h\nu = e^2/16\pi\varepsilon_0 r_0 (a^2 t^2/c^2)$

eliminating ν from above two equations

$t^4 = (2\epsilon_0 hc/e^2)(4r_0 c^2/a^3) = 1/\square (4r_0 c^2/a^3)$

$\square$ is fine structure constant, r_0 is the value of charge radius = $\square r_m$

r_m is mass radius of electron = $h/4\pi mc = 1.93 \times 10^{-13}$m

m is the mass of electron putting the above values $t^4 = 4r_m c^2/a^3$

t is the time interval required for formation of photon during acceleration a of electron putting the values of constants $t^4 = 7 \times 10^4/a^3$ metre^3 sec^{-2} a is the value of acceleration of electron

Evolution of Physical Laws

Some time period is required for the conversion of electromagneton energy to photon. Therefore, electron needs to be accelerated for that time. Uniform acceleration leads to accumulation of energy and prevents the emission of discrete photon. The electron needs to be accelerated in knee jerk fashion.

The above result shows that when the charged Akashon starting from rest acquires velocity v, the extra electrical potential energy given by the above equation is added up to the electromagneton.

The increase in electromagneton energy on being accelerated is changed to photon energy. So from above discussion we know that when the charge packet is accelerated, it is the excess electrical potential energy stored in the electromagneton on being set into motion that is changed into photon energy. So long as the charge packet moves with uniform velocity V, no electromagnetic radiation is emitted. When the electron is moving around the positive nucleus no radiation is emitted because velocity V of the electron does not change hence, electron will not fall into the nucleus and thus **the main objection to the Bohr concept of stationary orbits is ruled out.** Only specific velocities can form photon (see next). According to the classical concept, the cause of emission of electromagnetic energy when the charge is accelerated is the bending of unrealistic lines of force which is not so convincing and straight forward and also requires complicated mathematics and predict the continuous emission of radiation. The procedure adopted here is very clear and simple. The photon is also emitted if charge packet in motion is decelerated because fall in electromagnergy of electromagneton is changed to electromagnetic energy. Actually the value of ΔE calculated above is more than the real value because the approximation v << c is used, but the error introduced is very very less and can be easily neglected. The contribution of self gravitational potential energy as calculated later on is also neglected.

Self Elcetrostatic Potential Energy of Pure Charge Packet:

In previous calculations self electrical potential energy of hybrid charge was calculated. Suppose q is the pure charge present on charge packet. Suppose as the charge packet is being built up, r is the radius of sphere of tranquillity at any instant. The pure charge q is equal to $\sqrt{(hc\epsilon)}$, ϵ is the permittivity at a point located on the surface of sphere of tranquillity. Suppose additional charge dq is added to the pure charge packet, pure charge q is supposed to be located at the centre of sphere of tranquillity, while additional charge dq is added to the surface of sphere of tranquillity. Change in electrostatic potential energy dE is given by,

$dE = q.dq/4\pi\epsilon r$

The value of ϵ is the instantaneous value when the radius is r. The value of ϵ obeys the principle of inverse variation.

$4\pi r\epsilon = k_\epsilon$ = constant
$q = \sqrt{(hc\epsilon)}$
$dq = \frac{1}{2}\{hc/\sqrt{(hc\epsilon)}\}.d\epsilon$
$\epsilon = k_\epsilon/4\pi r$

Evolution of Physical Laws

$$dE = \{\sqrt{(hc\epsilon)}/k_\epsilon\}.\{(d\epsilon).hc/2\sqrt{(hc\epsilon)}\} = (hc/2k_\epsilon).d\epsilon$$

$$E = \int_0^E dE = hc/2k_\epsilon \int_\infty^r d\epsilon = (hc/2k_\epsilon).\epsilon = hc/8\pi r$$

The expression for energy is independent of ϵ.

How this self electrical potential energy of pure charge packet changes into photon is discussed in the next topic.

NATURE OF PHOTON AND PHOTON ENERGY:

Einstien remarked in a letter to M. Besso, "All these fifty years conscious brooding have brought me no near to the answer to the question 'What are light quanta?' Now a days every Tom, Dick and Harry thinks he knows but he is mistaken". It is very much true even today. It is to be judged how far I am successful in removing the veil of mystery regarding the nature of photon. The light (Prakasha) is regarded as special gift of nature which destroys ignorance or darkness, but it is the irony of nature that we are ignorant of nature of light (knowledge). How ignorance (darkness) and light (knowledge) can co-exist!

It has been seen that the photon is born when charged Akashon is accelerated. The energy packed up in photon is the excess energy of the temporary electromagneton associated with the accelerated charged Akashon. So it can be said that the photon is progenitor of temporary electromagneton of the charged Akashon, therefore, the nature of the photon is the same as that of electromagneton. The temporary electromagneton is combination of two energy packets carrying equal and opposite pure charges with oscillating charge shells. Therefore, the photon is composed of doublet of electromagnetons carrying equal and opposite pure charges. As will be shown later on that this photon is not so stable. It can be called as pro-photon. The nature of the charge is pure, therefore, each energy packet carries charge equal to $\sqrt{(hc\epsilon)}$, the charge being pure, the value of charge can change with change in ϵ. That photon carries doublet of opposite charges is born out by the fact that when photon passes near the nucleus, the nucleus charge breaks up the photon into two oppositely charged particles electron and positron.

Metamorphosis of Electromagneton Energy to Photon Energy (Birth of Photon):

When the charge packet (electron) is accelerated the radius of sphere of tranquillity r_0 decreases to r, hence during this transition pure charge is liberated. The pure liberated charge accommodates the excess self electrostatic potential energy evolved as a result of motion to form temporary electromagneton having pure charge. The singlet state of the pure charge elcetromagneton is highly unstable, hence it breaks up into two parts having equal and opposite charges to form neutral doublet composed of two pure electromagnetons when the pure charge shells of the doublet start oscillating, the doublet changes to photon.

$$\text{Doublet of electromagneton} \rightarrow \text{Photon}$$
$$(\pm) \qquad\qquad (\pm)$$

Suppose r' is the radius of sphere of tranquillity of one electromagneton.

As already deduced self electrostatic potential energy of pure charge packet or singlet electromagneton = $hc/8\pi r'$

Energy of doublet = $2hc/8\pi r' = hc/4\pi r'$

Extra self electrostatic potential energy liberated as a result of motion of hybrid charge = $e^2/8\pi\epsilon_0 r_0.(v^2/c^2)$

$hc/4\pi r' = e^2/8\pi\epsilon_0 r_0.(v^2/c^2)$

$r' = r_0.c^2/\alpha.v^2$

Evolution of Physical Laws

$\alpha = e^2/(2\epsilon_0 hc)$ = fine structure constant

$c > v$, $\alpha < 1$, $r' > r_0$

The photon is more bloated than the charge packet.

Energy of photon = E = Energy of doublet = $hc/4\pi r'$

$4\pi r'.E = hc$

$4\pi r'.\nu = c$

$E = h.\nu$

The above is Planck's relation for energy of photon.

The pure charge liberated is carried away by the photon hence there is no increase in the charge of the electron as a result of acceleration. During acceleration of electron inertial electromagneton is generated which is associated with pure charge. The tiny pure charge of the inertial electromagneton is divided into two halves; one half is carried away by the photon while other half is retained as magnetoelectromagneton. The inertial electromagnetong is responsible for the dual nature of photon.

The above is the best method for the birth of photon. The birth of photon can be explained in slightly modified version also.

It will be now seen how the extra temporary electromagneton formed as a result of acceleration changes to photon. It has been seen that the extra energy added up to the electromagneton as a result of motion leads to unstable state of electromagneton therefore the extra energy is radiated out as photon to acquire stability as soon as it is generated. The extra energy is present in the form of pro-photon. The metamorphosis of pro-photon to photon is three step process.

$$\Delta E = (e^2/8\pi r_0 \epsilon_0).(v^2/c^2)$$

Step 1:

In first step, pro-photon velocity becomes equal to c. Therefore, v = c so that energy of pro-photon may be equal to that of electromagneton, $r_0 \rightarrow r'$ where $r' > r_0$.

Step 2:

Frequency ν_0 is changed to pro-photon frequency ν'

$4\pi r'\nu' = c$

Energy of pro-photon $\Delta E = e^2\nu'/2\epsilon_0 c$

Step 3:

In this step pro-photon changes to photon when ν' changes to $\alpha'\nu$ where, ν is the frequency of photon, and α' is dimensionless constant.

Finally, the energy of photon $= \alpha' e^2 \nu / 2\epsilon_0 c$

Put $\alpha' e^2 / 2\epsilon_0 c = h = $ constant

Therefore, energy of photon $= h\nu$

ν is the frequency of photon

$\alpha' = 2h\epsilon_0 c/e^2 = 1/\alpha$

Here, α is fine structure constant.

The above is very simple proof of Planck's energy relation for photon.

During above steps radius of sphere of tranquillity of emitted photon is more than that of pro-photon or charged particle, say electron. Thus, photon is more bloated than electron. The source of photon energy is extra self electrical potential energy acquired as a result of acceleration of the electron or charged particle.

From above equation it is seen that photon may change to singlet state or doublet state.

$$\text{Doublet} \rightleftarrows \text{Singlet} \rightleftarrows \text{Singlet}$$
$$(\pm) \qquad\qquad (+) \qquad\qquad (-)$$
$$stable \qquad\quad unstable \qquad\quad unstable$$

That photon is doublet of charge packets is born by the fact that photon gives pair of positron and electron.

Photon → Positron + Electron

In above calculations ν is the frequency taken equal to the frequency on the surface of sphere of tranquillity but actually ν has got many values. Dominant value of ν is on the surface of sphere of tranquillity because other values of ν falls rapidly with distance.

$d\nu/dr = -(c/4\pi r^2)$

Because of large value of c and small value of r, rate of change of frequency with distance is very large.

In addition to the permanent charges on doublet, photon also carries tiny pure residual charge which is discussed later on. When the photon moves from one place to another the motion can be equated to the motion of tiny residual free charge packet but with the difference, the sign of the charge is changing in sinusoidal fashion, therefore, the electric field associated will also trace same curve. Because the charge is moving therefore, the magnetic field will also be there. The magnetic field will also be changing in the same manner as the electric field. The combination of two fields form changing electromagnetic field. The final conclusion is that changing electromagnetic field is associated with the photon.

When two singlet states of photon in opposite phase combine, there is destructive interference while there is reinforcement when the phase is same. The above observation shows that the interference associated with light radiation can be explained without envoking the wave like nature.

Electron as an SHO: When electron acting as SHO reaches the mean position the velocity is maximum and hence the maximum energy is stored in the temporary electromagneton. At the mean position half of the energy of temporaryelectromagneton is radiated while the other half remains associated with the electron to push it to the extreme position. After reaching the extreme position the electron again starts moving towards the mean position and on reaching the mean position again photon is ejected which shows that during one oscillation two photons are ejected. Suppose ν_0 is frequency of oscillation of SHO. The time interval between emission of two photons is $1/2\nu_0$. When the photon is ejected at mean position wave train is also associated with photon. Two photons which are emitted successively must be separated by half wave length so that wave train is continuous. Distance travelled by photon in $1/2v_0 = \lambda/2$. Distance travelled by photon in unit time = $v_0 \lambda$ = speed of light If vis is the frequency of radiation, speed of light = $v\lambda$ therefore $v\lambda = v_0\lambda$ which proves that $v=v_0$ or the frequency of radiation is equal to frequency of oscillation of electron. In time interval $1/2v_0$ two photons are emitted having energy = $2hv_0$ therefore, energy emitted per unit time = $2hv_0^2$.

Another form of electromagneton energy: Charge radius and mass radius of charged akashon are not equal. Suppose charge radius is r_0 and mass radius is r_m.

The ratio of two radii is $r_0/r_m = \alpha$

r_m = mass radius = $h/4\pi mc$

The energy of permanent electromagneton of charged akashon = $e^2/8\pi \epsilon_0 r_0$

Putting value of charge radius the electromagnerton energy = $e^2/8\pi\alpha r_m$

The electromagneton and mass energy are same. Mass energy changes to energy through electromagneton energy. Mass energy y= mc^2. Equating the mass energy and electromagneton $\alpha = e^2/2hc\epsilon_0$ = fine structure constant. While equating the mass energy and electron magneton energy the contribution of mattenergon being very small is neglected. Alternatively it can be seen that ratio of charge and mass radii is dimensionless constant and hence putting its value equal fine structure constant it can concluded that electromagneton energy = mc^2. It is thus SEEN THAT WHAT IS MASS OR MASS ENERGY IS SELF ELECTROSTATIC POTENTIAL

ENERGY OF HYBRID ELECTROMAGNETON, THEREFORE THE CAUSE OF MASS IS NOT THE HIGG'S BOSON ALL THE PARTICLES HAVE GOT SELF ELECTROSTATIC POTENTIAL ENERGY WHIC IS MANIFESTED AS MASS THE NATURE OF MASS IS NOT AN ENIGMA. The value of α shows that mass radius is greater than charge radius because α is less than one.

The energy of temporary electromagnton = $(e^2/8\pi\epsilon_0 r_0)v^2/c^2$
Put $r_0 = \alpha r_m$ and $r_m = h/4\pi mc$
Inserting the value of α and r_m, it can be easily seen that,
Energy of temporary electrromagneton = mv^2 = twice the kinectic energy
Total energy of temporary electromagneton usually exists as doublet therefore kinetic energy=energy of singlet state.

The change in velocity of electron leads to change in electromagneton energy and half of the change in electromagneton energy is radiated out as electromagnetic radiation. The cause of emission of electromagnetic radiation is not the bizarre concept of bending of non-existent lines of force. The energy radiated out is the self electrostatic potential energy of electron. It is also seen that mass energy changes to energy through electromagneton. The mass to energy change process is not direct process. This point will be discussed laster on also.

Planck's energy relation:
Electromagnetic radiation energy = $\Delta E = 8\pi\alpha$
ΔE is energy of photon of radius r_p

Therefore, $4\pi\Delta E = hc$

From above two equations it can be easily seen that $r_p/r_m = 2c^2/v^2$

So radius of photon is more than that of electron which emits photon. Energy of photon can be easily calculated by using the relation $r_p/r_m = 2c^2/v^2$ and putting value of α in the energy relation of electromagneton energy of photon $= hc/4\pi r_p = hc/\lambda = h\nu$. Thus Planck's energy relation is easily proved.

How mass changes to energy: The conventional mass packet is charged and has got two radii mass radius and charge radius. The ratio of two radii is equal to fine structure constant as already proved.

According to inverse variation law $4\pi r_m m = h/c$ and $4\pi r_e E = hc$, r_m is mass radius and r_e is charge radius because as already seen photon(energy) is charge packet with oscillating anti akasha shells.

When mass packet changes to energy packet $r_m = r_e$

by dividing the above two equations $E = mc^2$

It is thus seen that condition for change of mass into energy is that mass radius should become equal to charge radius and it happens so when electron and positron interact to form two photons because each one of the interacting particles changes to photon.

What is mass: Energy of permanent electromagneton of charged akashon (electron) at rest = $e^2/8\pi\varepsilon_0 r_0$. Suppose r_0 and r_m are the charge and mass radius of akashon(electron) at rest $r_0/r_m = \alpha$ = dimensionless constant.

Self electrostatic potential energy = $E_0 = e^2/8\pi\alpha\varepsilon_0 r_m$.
α is fine structure constant = $e^2/2\varepsilon_0 hc$
Put value of fine structutre constant $E_0 = hc/4\pi r_m$
$h/4\pi r_m = m_0$
$E_0 = m_0 c^2$

The above relation shows that self electrostatic potential energy and mass of an akashon are equivalent.

Self electrostatic potential energy ↔ Mass energy equivalent

Both are intercovertible. It is thus proved that mass is congealed form of self electrostatic potential energy. It is not the Higg's field or Higg's boson that impart mass to particle. IT IS THE SELF ELECTROSTATIC POTENTIAL POTENTIAL ENERGY WHICH IS MANIFESTED AS MASS, MASS CHANNGES TO ENERGY THROUGH ELECTROMAGNETON AND VICE VERSA

Quantization of electromagnetic energy: It has been seen that temporary electromagneton is radiated out as electromagnetic energy = $(e^2/8\pi\varepsilon_0 r_0^\wedge)v^2/c^2$
$r_0^\wedge$ is radius of sphere of tranquillity of electron charge

Suppose rest radius of electron is r_0 and r the radius in motion. As already seen $r = r_0\sqrt{1 - v^2/c^2}$ and from it $v^2/c^2 = (1-r^2/r_0^2)$ $r = na_0$
and $r_0\hat{} = n_0\hat{}\ a_0$, a_0 is the minimum measureable length.

Using the above relation, magnitude of radiation emitted by electron
$= (e^2/8\pi\epsilon_0\ n\hat{}_0 a_0)(1-n^2/n_0^2) = (e^2/8\pi\epsilon_0\ n_0\hat{} a_0)\{(n_0-n)((n_0+n)\}/n_0^2$
$n_0 - n = N$,
here $N = 1, 2, 3, \text{---------}$
therefore, emitted radiation energy $= \{N(2n_0-N)/n_0^2\}(e^2/8\pi\epsilon_0\ a_0 n_0\hat{})$
$n_0 \gg N$, therefore, $(2n_0-N) \approx 2n_0$
energy radiated $=(N/n_0)(\alpha hc/4\pi a_0 n_0\hat{})$ α is fine structure constant.
The value of emitted energy is $= N \cdot E_K$
$N = 1, 2, 3, \ldots$ and $E_K = \alpha hc/2\pi = $ constant

The following equation $\alpha c/a_0 n_0 n_0\hat{} = $ constant and has dimension of frequency $= v_0$; v_0 is unique natural frequency of cosmic symphony or this is the frequency of primordial disturbance after the birth of universe. Emitted radiation energy $= hNv_0 = h\square$; $\square = Nv_0$; $\square$ is the frequency of emitted photon which shows that photon frequency is integral multiple of basic frequency. The Plancks energy relation is thus easily proved. The value of v_0 for electron can be calculated byputting the values of constants.
Mass radius of electron $= 1.93 \times 10^{-13}$ m
Minimum length $= 1.14 \times 10^{-35}$ m
$n_0 = $ mass radius/minimum length $= 1.67 \times 10^{22}$
$n_0\hat{} = \alpha n_0$

$n_0 = 1.2 \times 10^{20}$

$a_0 = 1,14 \times 10^{-35}$

$c = 3 \times 10^8$ putting the values of constants

$v_0 = 1/10$ (approx.)

It shows that value of natural cosmic frequency is very small but not zero or everything in universe is vibrating. The minimumenergy of photon=hxminimum frequency=h/10/sec. very very small value. It can be said that nearly ten natural oscillations are equal to one conventional second. The natural clock is very slow.

THE BIRTH OF PHOTON AND ENERGY OF PHOTON:

This method is not so logical as the previous treatment. The previous methods discussed are the best methods to discuss birth of photon. However, the following procedure may also be discussed. The temporary inwards growth of the region of perturbation of the charge packet soon starts melting away because the extra self electrical potential energy stored in the temporary region of perturbation is carried away by the waves which start from the numerous oscillating shells located on the inner surface of the extra temporary region of perturbation. The wavelets move towards the centre of charge packet, the whole of the sphere of tranquillity is filled up with energy and extra temporary region of perturbation of the charge packet disappears. It is against the nature of sphere of tranquillity to retain disturbance for long, therefore, it restores its tranquillity by spilling out the energy filled bowl of the sphere of tranquillity with the ejection of packet of energy called Prakashon or photon. Let us see how it happens. The carriers of energy inside the sphere of tranquillity are the electromagnergy waves which are pro electromagnetic waves. There is some incubation period for the formation of Prakashon (photon). The center of the pro-photonwhich is associated with electromagnergy waves overlap the center of the sphere of tranquillity of the charge packet. The pro-Photon center does not move along with the center of Akashon hence, velocity of light is independent of source velocity. If v is the velocity of the Akashon at any instant, the velocity of matter waves is c^2/v as will be seen later on. The velocity of electromagnergy waves is $\alpha'c^2/v$, where α' is dimensionless constant. The formation of next photon starts after the charged Akashon has traveled distance equal to the radius of the sphere of tranquillity of the charge packetbecause the two centers of pro-photon cannot be simultaneously located in the sphere of trqnauillity of the charged Akashon. The above observation shows that emission of photon is not continuous process. If v is the velocity acquired after acceleration and r is the radius of sphere of tranquillity respectively, the time required to cover distance equal to r_0 is r_0/v. The charged Akashonis subjected to acceleration by supplying force in knee jerk fashion therefore, time spent in acquiring velocity v can be neglected. It is the time interval between shooting out of two successive photons. If the Akashon is having uniform acceleration there is no emission of photon because no incubation time is there for the formation of photon. The distance traveled by the electromagnergy waves during the time interval r_0/v is equal to the radius of sphere of tranquillity R of the photon which is to be ejected out after calculated time interval iscalculated below:

$$\alpha'.(c^2/v).(r_0/v) = \alpha'.(c^2/v^2) \, r_0 = R$$

The energy carried by the outward flow of the electromattenergy (pro-electromagnetic energy) waves starting from the oscillating charge center gather into energy packet called Prakashon or photon. The distance which is calculated above is the radius of sphere of tranquillity R of the ejected photon. Let us next see how much electromagnetic energy is packed up in Prakashon (photon).

Total energy = $(e^2/8\pi\in r_0).(v^2/c^2)$

As already calculated,

total energy associated with the photon = $e^2/8\pi\epsilon_0 r_0 \cdot (v^2/c^2)$ because it is the extra electromagnergy of the charge packet which is changed into photon energy.

The radius of sphere of tranquillity of photon = $R = \alpha' \cdot (c^2/v^2) r_0$
$\Rightarrow v^2/c^2 = \alpha'(r_0/R)$

Energy packed up in photon = $(e^2/8\pi\epsilon_0 r_0)(v^2/c^2)$
Put the value of (v^2/c^2)
The energy of photon = $(e^2/8\pi\epsilon_0 r_0)\alpha' r_0/R = \alpha' e^2/8\pi\epsilon_0 R$

R is the radius of sphere of tranquillity of the emitted photon
$4\pi R\nu = c$
$R = c/4\pi\nu$

Put the value of R,

Energy of photon = $\alpha' \cdot (e^2/2\epsilon_0 c) \cdot \nu = K\nu$, where $K = \alpha' \cdot (e^2/2\epsilon_0 c)$, K is constant. The above result proves that, energy of photon is proportional to frequency.

So farthere has been no comprehensive derivation which proves that energy of photon is proportional to frequency.

Put K = h (Planck's Constant)

$h = \alpha'.(e^2/2\epsilon_0 c)$

therefore, $\alpha' = 2\epsilon_0 ch/e^2$

or $\alpha' = 1/\alpha$

α is fine structure constant

The above relation proves the famous Planck's relation for the energy of photon which was taken as suchwithout any comprehensive derivation for the energy of photon.

The above discussion shows that the energy radiated out is not continuous but as discrete packets of energy.

As already discussed, though the Akashon continues to move, the source of matter charge waves is the stationary oscillating point and after the filling up of the sphere of tranquillity with energy, the energy is carried away by Prakashon or photon which moves with constant velocity C relative to the stationary point which shows that the velocity of light is independent of velocity of source. The birth of the Prakashon can be visualized as:

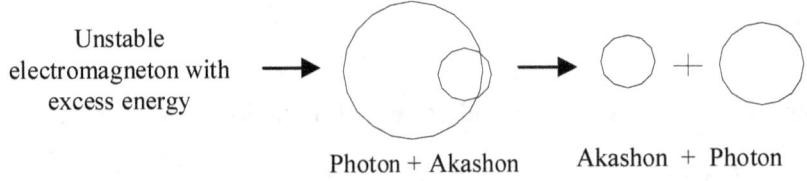

Unstable electromagneton with excess energy → Photon + Akashon → Akashon + Photon

Fig. 29

The photon formed is separated from the electron by the weak or separating force as discussed during the breaking up of unstable elementary particles to form smaller particles.

That the velocity of light is independent of the source velocity can also be explained in another manner. During the acceleration of charged Akashon, energy liberated is used in forming positive or negatively charged electromagneton. The pro-photon first formed from the excess electromagneton energy is dismantled and reforms itself by using shells of free anti-ether present all around to form photon. The photon separates from the electron in the same manner as discussed above by using weak force or separating force. Because the photon is formed by using shells of the free ether which are stationary, therefore, velocity of photon or light is w.r.t. the stationary shells and hence, is independent of the velocity of electron which is source of photon or light. During special theory of relativity, the concept of absolute frame of reference was discarded. I reintroduce the concept. The absolute frame of reference is the frame containing stationary shells or free ether or Akasha. The velocity of light is always w.r.t. this frame whether the source is moving or not.

ASSOCIATION OF ELECTRIC AND MAGNETIC FIELDS WITH PHOTON:

It was seen that when the photon is emitted from a charged source some tiny residual charge is liberated which is associated with the photon. It was also seen that in doublet state two energy packets or electromagnetons carrying opposite charges are present. The charge on these energy packets is not equal at any instant because a very small amount of tiny residual charge formed as a result of association of inertial electromagneton with the photon is associated alternatively with each part of the doublet of energy charge packets or the electromagneton in very quick succession. The regions of perturbation of two energy packets undergo change in out of phase oscillationary mode i.e. if residual positive charge on one energy packet is increasing the residual negative charge on other energy packet is decreasing. The radius of sphere of trqnauillity of the permanent region of perturbation of each of the energy packet remains same; it is only a little extension of the region of perturbation that leads to the formation of oscillating temporary region of perturbationhaving dual nature as explained earlier.The temporary regionof perturbation is associated with ministream amount of energy as compared to the energy of permanent region of perturbation.

Suppose first residual positive charge increases and residual negative charge decreases on two temporary regions of perturbation. Therefore,at a certain stage free residual positive and negative charges will be equal and hence net residual charge will be zero, further on negative residual charge starts increasing and hence positive charge starts decreasing.At this stage, there is net negative charge, therefore, sign of the field is negative. After some time, the negative charge is maximum and positive charge is zero, therefore, negative field is maximum. Further on negative charge starts decreasing and positive charge starts increasing. After some time opposite charges will be equal and hence electric field drops from maximum to zero, and the same cycle is repeated with net residual positive charge. It is seen that it is the tiny residual charges of opposite types associated with each electromagneton which is the cause of sinusoidal variation of net charge and hence sinusoidal electric field is created. Because the tiny sinusoidal residual charge is moving with the photon, therefore, sinusoidal magnetic field is also associated with the photon.

Electromagnetic waves are studied disregarding the particle like nature, but in the present text focus is on the particle nature because internal structural arrangement of the photon is known for the first time and hence simultaneous particle like nature and wave like nature can be explained. Two permanent electromagnetons associated with the photon impart particle like nature while, temporary regions of perturbation formed as a result of tiny residual pure charge associated with the electromagnetons is the cause of wave like properties of the photon. While discussing the electromagnetic waves presence of free charge is ignored by disregarding the basic fact that no electric or magnetic fields are there without any free charge. In the above discussion it has been seen that tiny oscillating net charge continuously changes sign, which is the cause of oscillating electric and magnetic fields. The charge q, Electric field and magnetic fields are represented by the sinusoidal equations as given below:

$q = q_0 . \sin(\omega t - kr)$

$E(rt) = E_0 \sin(\omega t - kr)$

$B(rt) = B_0 \sin(\omega t - kr)$

Suppose two pure charge packets are moving parallel to each other with velocity of light, pure charge packet, means that no mass is associated with the charge packet. The similar charge packets moving parallel to each other in the same direction experience repulsive force due to electrostatic force and attractive force due to magnetic force. No extra force should act on the charge packet because it is already moving with the limiting velocity c. Therefore, the two forces should be equal.

F_1 = Electrostatic force = $(1/4\pi \epsilon_0)(q_1 q_2 / r^2)$

F_2 = Magnetic force = $(\mu_0 / 4\pi)(q_1 q_2 c^2 / r^2)$

$F_1 = F_2$

Therefore, $\mu_0 \epsilon_0 = 1/c^2$

The transitory or residual charge developed on the Prakashon moves with the velocity of light, therefore, the electric and the magnetic fields are equal at any instant. As already seen the electric and magnetic fields are oscillating.

Electric field = $E = E_0 \cos(\omega t - kr)$

Magnetic field = H

$E = H$, $H' = Bq.c$

q is the residual charge at any instant

$H = H'/q = Bc$

$E = H = Bc = E_0 \cos(\omega t - kr)$

$B = E_0 \cos(\omega t - kr)/c = B_0 E_0 \cos(\omega t - kr)$

$E/B = E_0/B_0 = c$

The above relation has been derived without using complicated mathematic calculation. E is much greater than B. The photon can be visualized as given below:

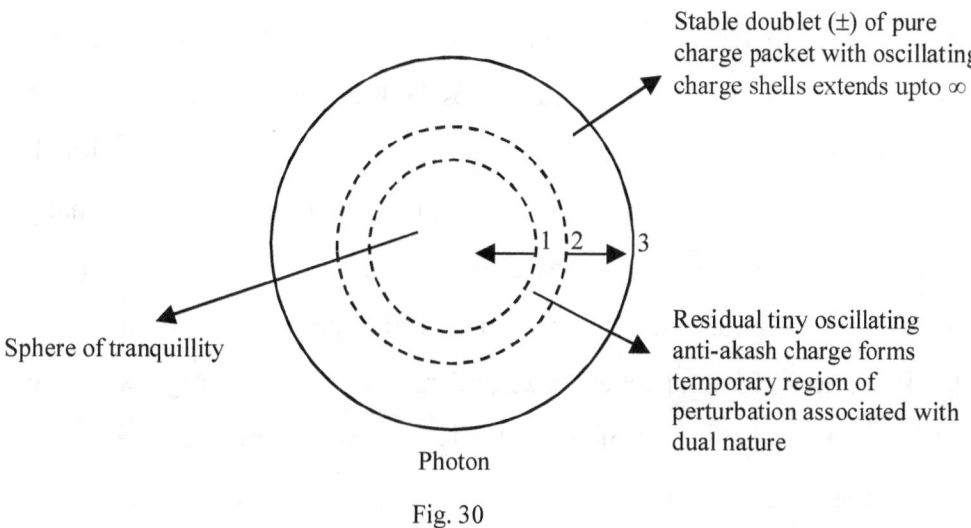

Fig. 30

1. Increasing boundary of temporary region of perturbation of energy packet 1 (+) residual charge.
2. Decreasing boundary of temporary region of perturbation of energy packet 2 (-) residual charge.
3. Boundary of two permanent regions of perturbation of two energy packets ($\pm$) residual charge.

Dual Nature of Photon:

The particle like nature is exhibited by permanent regions of perturbation and wave like nature is associated with the temporary region of perturbation which is composed of bonded and free anti-ether, hence exhibits dual character. The dual nature will be also explained in detail later on.

The Velocity of Light and Source Velocity:

It can be shown easily that the velocity of photon is independent of the velocity of source i.e. accelerated charge. Suppose photon has emitted in the direction of motion of charge packet, if the velocity of the photon is affected by the source velocity, the relative velocity of the photon should by $(c - v)$, while if it is emitted in the opposite direction the relative velocity value will be $(c+v)$. The change in the velocity of the photon with the change in the direction of emission will change frequency and hence change in the energy of photon but that is fixed. Therefore, the conclusion is this that velocity of photon is independent of the direction of emission or velocity of the charge packet because only then the energy carried by photon will be independent of the velocity of source. This statement is the corner stone of the special theory of relativity. The photon is emitted in a direction perpendicular to the direction of motion of charged packet. We have seen here that it has logical justification. This very property of the photon shows that Michelson-Morley experiment is not suitable experiment to detect the presence of ether.

THE SELF GRAVITATIONAL POTENTIAL ENERGY OF PURE MASS AKASHON AND HYBRID AKASHON:

The gravitational force like the electrostatic force can be expressed as:

$((1/(4\pi G')).(M_1.M_2))/r^2)$

$G = ¼(\pi G')$

Self gravitational potential energy can be calculated in the same manner as electrostatic potential energy by changing (epsilon) to G' and e to m. The result for self gravitational potential energy can be got from self electrostatic potential energy deduced earlier by making above changes.

Self gravitational potential energy of hybrid Akasha = mattenergon = $(Gm_0^2/2).(1/r_0)$ While driving the above relation G' varies according to $4\pi rG' = k_g$ = constant

Pure charge is expressed as $\sqrt{(hc\epsilon)}$. As already proved pure mass can be expressed as $\sqrt{(hc/4\pi G)}$. In case of pure charge packet, the value of G varies as $4\pi rG^{-1} = k_g$. This relation has already been proved G_p is the value on the surface of sphere of tranquility of radius r_p. Further on notation p can be dropped.

Potential energy $dE = -G.m.dm/r$

The above is the change in potential energy as region of perturbation is being formed by the addition of more and more pure mass.

$$m = 1/4\pi(\sqrt{\square ck_G}/r)$$

$$dm = -1/8\pi(\sqrt{\square ck_G}) \cdot r^{-3/2} dr$$

Self gravitational potential energy $= E = hc/8\pi \int_r^\infty dr/r^2 = -hc/8\pi r$.

It is thus seen that self gravitational potential energy of pure mass mattenergon is equal self electrostatic potential energy of pure electromagneton.

The above energy is packed as mattenergon in the same manner as energy of electromagneton.

Pure electromagneton energy $= (e^2 / 8\pi \in_0) \cdot (1 / r_0)$

Pure mattenergon energy $= G \cdot m_0^2 / 2r_0$

Ratio of the two energies $= e^2 / 4\pi G m_0^2 \in_0 = 1.7 \times 10^{33}$

This is huge energy difference in favour of electromagneton.

The very small value of energy associated with mattenergon makes it very difficult to detect gravitational waves if these are formed at all.

In case of electromagneton the anti-Akasha shells are stretched to limit therefore the excess electromagneton energy acquired is radiated as photon, but in case of mattenergon, the excess energy does not go out as graviton. It is used to create temporary region of perturbation. It is only in case of super heavy Akashon which will be examined later on that the excess energy acquired as a result of motion is radiated as graviton, therefore, acceleration of super heavy Akashon is good source of gravitons. The gravitons can also be produced on deceleration of massive Akashon because here energy stored in the temporary region of perturbation is good source of gravitons. It is concluded that emission of gravitons and photons are not exactly similar.

The energy of the electromagneton and mattenergon will be considered later on while bringing about **unification of gravitational and electromagnetic fields.**

Evolution of Physical Laws

THE UNIFICATION OF ELECTROMAGNETIC FIELD AND GRAVITATIONAL FIELD AND NATURE OF PHOTON:

The unification of electromagnetic and gravitational fields can be explained in another way also.

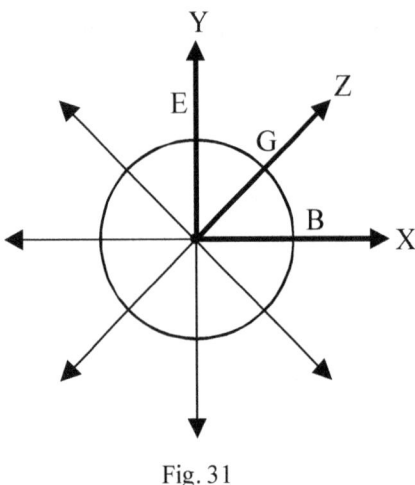

Fig. 31

Electromagneton (electromagnetic field) + Mattenergon (gravitational field) = Photon (fields unified)

Pure electromagneton: Pure electromagneton is composed of pure charge shells.Electromagneton has got sphere of tranquillity and region of perturbation. When it is composed of only one type of antiakasha shells its state is singletThe singlet state combines with other singlet state carrying opposite charge to form stable neutral doublet(±). It will be shown that photon is doublet of electropmagnetons or charge packets Self electrtostatic potential energy is stored in the electromagneton.

Pure mattenergon: Pure mattenergon is composed of pure mass shells. Singlet mattenergon is composed of negative or positive mass shells Stable doublet mattenergon (±) is composed of opposite types of mass shells. Self gravitational potential energy is packed up in mattenergon.

Electrtomagnetic field is associated with electromagneton and mattenergon carries gravitational field If r is the radius of sphere of tranquillity of pure electromahneton and pure mattenergon energy associated with each is equal to $hc/8\pi r$. The two are indistinguishable from energy point of view. Electromagneton<-- -> Mattenergon. It shows that electromagnetic and gravitational fields are inter conconvertible or thus are unified through electromagneton and mattenergon

The photon is composed of electromagneton and mattenergon, therefore, energy of photon is shared by both The notation for photon is $A^{\wedge}(\pm).A(\pm)$ electromagnetic field is associated with electromagneton while gravitational field is associated with mattenergon hence photon unifies electromagnetic and gravitational fields (see diagram). The proposed structure of photon is justified by pair production and annihilation of particles. The gravitational bending of light is also strong proof.

Interaction of Particles: There are two mass shells $A(+)$ and $A(-)$, similarily there are two charge shells $A^{\wedge}(+)$ and $A^{\wedge}(-)$. The following combinations are possible $A^{\wedge}(+)A(+),A^{\wedge}(-)A(-),A^{\wedge}(+)A(-)$ and $A^{\wedge}(-)A(+)$. Out of these only two combinations formed by pairs of opposite charge and mass shells are stable The stable pairs are composed of hybrid charge-mass shells, the building blocks of conventional matter. The notation for electron is $A^{\wedge}(-)A(+)$ and for positron $A^{\wedge}(+)A(-)$.

Pair production: Photon -------> electron + positron

In terms of notation,

$A^{\wedge}(\pm)A(\pm)$ (photon) ------> $A^{\wedge}(\pm)$(electromagneton)+$A(\pm)$(mattenergon)

If photon energy is equally shared by electromagneton and mattenergon through exchange of shells electron and positron are formed

$A^{\wedge}(\pm)A(\pm)$ ----------> $A^{\wedge}(-)A(+)$[Electron]+$A^{\wedge}(+)A(-)$[Positron].

Annihilation: In annihilation process electron+positron -----> two photon.

The annihilation can be explained using notations as,
$A^{\wedge}(-)A(+)$(electron)+$A^{\wedge}(+)A(-)$(positron) ------->
$A(\pm)$(electromagneton)+$A(\pm)$(mattenergon) ------->
$A^{\wedge}(\pm)A(\pm)$[photon]+$A^{\wedge}(\pm)A(\pm)$[photon]

The electromagneton and mattenergon produced are in the singlet state of photon where energy of photon is concentrated in electromagneton or mattenergon The singlet state is unstable and hence each of the electrimagneton and mattenergon changes to stable state where energy of photon is distributed over electromagneton and mattenergon and thus two photons are produced It is also seen that when electron and positron interact the charge and mass radii of electron and positron become equal hence electron and positron change to two photonsThe hybrid akashon or conventional mass packet changes to photon when charge radius and mass radius become equal .Before annihilation electrostatic ,magnetic and gravitational fields were associated with electron and positron therefore ,photon is carrier of all the three fields Again it is proved that photon unifies electromagnetic and gravitational fields .Exact bending of light as calculated by using CLASSICAL MECHANICS as discussed in general theory of relativity prove the composition of photon as discussed above

.

A^($\pm$)+A($\pm$)(photon)--------> A^($-$)A($+$)(electron)+A^($+$)A($-$)(positron) in the pair production electron and positron produced are in state of motion, therefore each carries electrostatic ,magnetic and gravitational field but if photon as, according to current practice is considered to be associated with only electromagnetic field ,the common sense demands that gravitational field should also be associated with photon which is transferred to electron and positron on pair production The simple discussion leads to the important conclusion that electromagnetic and gravitational are unified in photon as mattenergon and electromagneton The pair production supports proposed composition of photon.

Exact bending of light by classical mechanics: The bending of light by classical mechanics also supports the proposed composition of photon .Soldner failed to arrive at the exact solution because he did know the composition of photon.The eccentricity of path followed by mass m in gravitating body of mass M is given below,

Eccentricity $e = \sqrt{1 + 2EL^2/G^2 m^3 M}$

E and L are energy and angular momentum.

Neglecting gravitational energy $E=1/2mv^2$, $L=mvR$

R is the distance of mass m from gravitating body. In the presence of strong gravitational field the energy of the photon is equally divided between electromagneton and mattenergon hence mattenergon which experiences gravitational field is associated with half of the mass 1/2m electromagneton does not contributes towards the gravitational force because it interacts only with the electrostatic field(m is the mass equivalent of photon). Due to above effect the angular momentum is reduced to half, $L=1/2mvR$ while calculating the energy mass m is not halved because mere bending of light path does not affect energy. The value of energy is the same before and after bending of path in case of photon v=c hence speed is very high therefore neglecting 1 in the bracket $e \approx c^2R/2GM$ angle of deflection=π-$2\cos^{-1}(1/e)$, hence $\cos^{-1} 1/e = \pi/2 - 1/e$. Using the above relations angle of deflection $\delta = 4GM/c^2R$. The above value of angle of deflection calculated by using CLASSICAL MECHANICS is exactly equal to value calculated by applying general theory of relativity using complex mathematics. The exact value of angle of deflectin strongly supports the proposed composition of photon. So far it has been said that photon is associated electromagnetic field only(electrostatic+magnetic) but now it is seen that photon is very much carrier of gravitational field also hence photon is agent for unification of all the three fields. Einstein's dream was to derive two inch long equation for the unification of fields but here it is seen that experimental results conclusively prove the unification process through photon. The photon was an enigma for Einstein. The composition of photon suggests that light is expected to bend in strong electic field also.

THE PROOFS IN FAVOUR OF THE NEW MODEL OF PHOTON:

1. **Compton Scattering:** Photon-electron exchange energy, it is not always that whole of the energy of the photon is taken up by the electron. Sometimes, only small chunk of energy of the photon is exchanged. This is called Compton Scattering. Let it be seen how exchange of energy in chunks is possible. There is no logical answer if the nature of photon as known till today is used to explain Compton Scattering. But according to the new model of photon and electron as present in this text the answer is not difficult to find out. It is known that photon contains doublet packet of electromagneton filled with electrical potential energy. Electron also contains elctromagneton along with mattenergon. It is thus seen that photon is mostly filled with self-electrical potential energy and some self electrical potential energy is also associated with electron aselectromagneton. It is through the exchange of this electrical potential energy either partially or fully that Compton Scattering or fully absorption of photon is possible.

2. **Pair Production:** It is known that when photon having energy at least equal to the twice the energy of electron passes very close to heavy charged nucleus, the photon splits into electron and positron. The photon consists of two charged packets carrying equal and opposite charges. The charge on the nucleus splits the two charged packets apart. The electromagneton energy which is associated with photon undergoes metamorphosis into mass energy of electron and positron.

3. **Annihilation of anti-particles:** Electron and positron produce two photons on interaction. Here reverse is the case as compared to above phenomenon. Some electrical potential energy is associated with the electron and positron before interaction as electromagnetons. When the anti-particles interact, the mass energy of electron and positron changes into two packets of electromagneton for association with two opposite charge packets supplied by oppositely charged anti-particles. The big photon so formed splits into two small photons to conserve momentum.

Evolution of Physical Laws

INTERACTION OF PARTICLES:

The interactions which are being considered are in favour of the shell structure of the Akashon and photon hence proves the presence of Akasha. The electromagneton and mattenergon are real particles unlike the QED virtual particles. These particles play important part in particle interaction. The interaction also proves the proposed composition of photon.

(1) **Evolution of photons during acceleration of charged packets:** This interaction has been discussed in detail. The additional temporary electromagneton is ejected as photon.

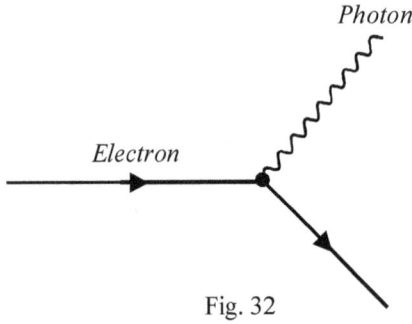

Fig. 32

(2) **Absorption of photon by charged Akashons:** The photon is absorbed to create additional temporary electromagneton. The photon is absorbed by the charged particle when the residual oscillating charge on the photon is of opposite sign to that of the charged particle. The photon center lies in the sphere of tranqulillity of charged particle hence state of charged particle becomes unstable. The photon is expelled out so that charged particle may become stable.

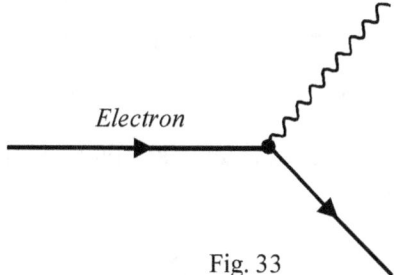

Fig. 33

The emission and absorption of photon can be explained by using concept of electromagneton.

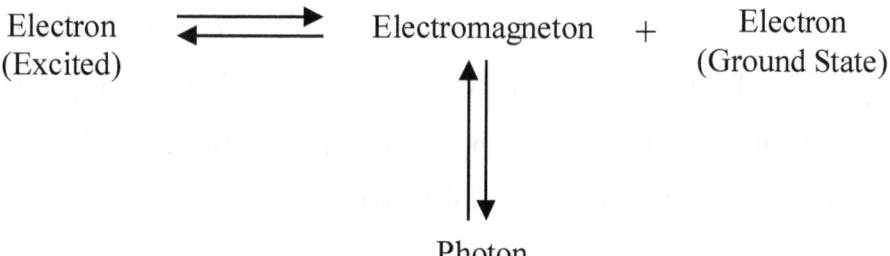

(3) Pair production:

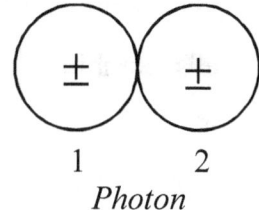

Photon

1. Mttenergon
2. Electromagneton

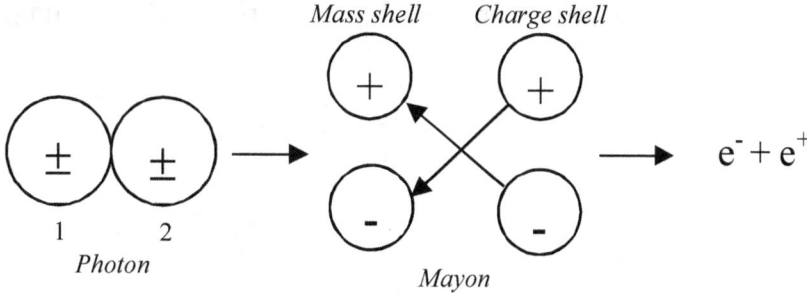

The nature of Akasha or anti-Akasha shells of which mass pakets or energy packets are composed of need to be discussed nefore discussing the topic. The Akasha and anti-Akasha shells are interconvertible.

Akasha shell ⇌ anti-Akasha shell

Pure charge shell is unstable. The stable state is achieved when it combines with pure mass shells to form hybrid mass charge shell. It also combines with pure charge shell of opposite size to form doublet state. When energetic photon passes near the charged nucleus the positive charge on the nucleus attracts the negative component of the electrical doublet of photon hence overlapping positive and negative charge centers of electrical doublet are separated.

The separation of charge centers expel the charge packets to form two pure charge packets as already mentioned pure charge shells are unstable. The pure charge shells combine with mass shells to form hybrid mass charge packets or one electron or one psitron. The pair production can be explained through electromagneton and mattenergon. The photon as it passes near the positive nucleus is split into pair of oppositely charged electromagneton. The electromagnetons further change to charged mattenergons which give rise to electrn and positron.

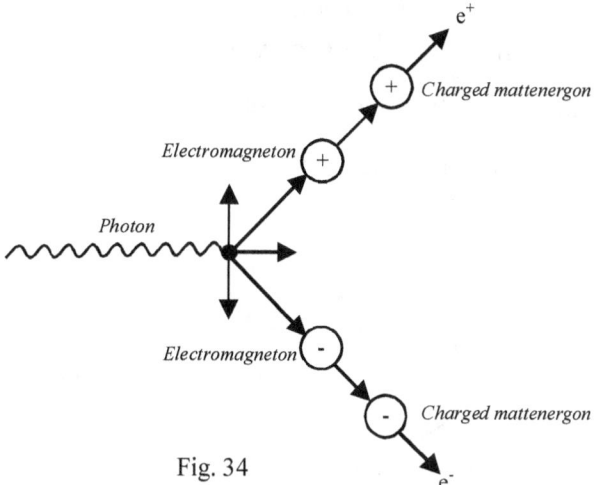

Fig. 34

(4) Annihilation of electron and positron pair: The annihilation is reverse of pair production as visualized in the diagram.

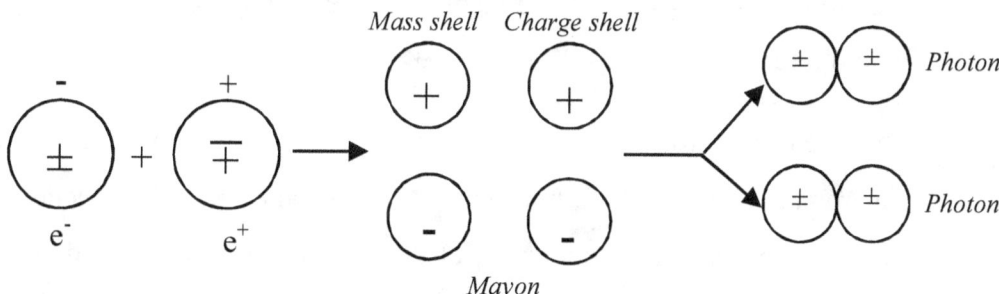

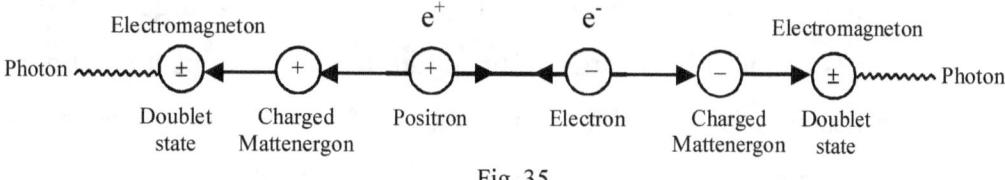

Fig. 35

(5) **Neutrino:** When three shells out of which one is positive or negative Akasha shell and other two are doublet of positive and negative charge shells. This triad is called neutrino. It is neutral. Neutrino is of two types.

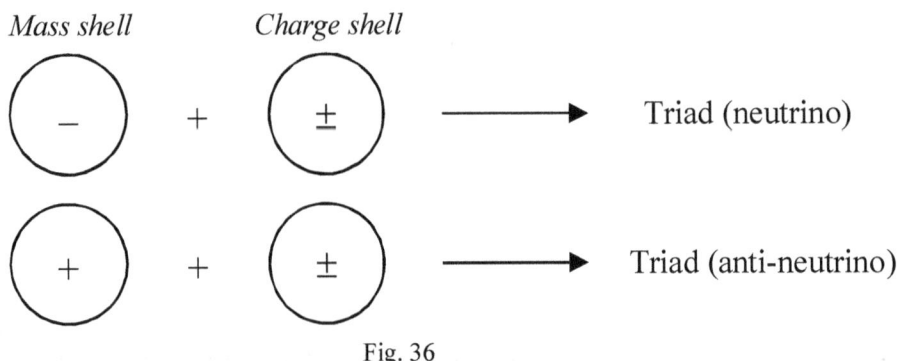

Fig. 36

The neutrino is neutral and very stable.

Neutron ⟶ Proton + Electron + Neutrino
Neutron ⟶ Proton + Electron (electromagneton)
Electron (electromagneton) ⟶ Electron + Neutrino

The electron formed in the first stage is associated with electromagneton which forms triads of shells called neutrino. Neutrino carries away the excess energy of electron. Doublet of charge shells in the formation of triad is supplied by electromagneton.

When anti-neutrino interacts with proton, positive charge of proton extracts the negative charge anti-neutrino to change to neutron and the anti-neutrino is left with hybrid shells of positive mass shells and positive charge shells to form positron. The interaction of charge shells is not easy because of stability of triad hence most of the neutrinos move on without any interaction.

As seen above in the five interactions, the presence of free ether shell is necessary. The interactions have been discussed very easily and also vindicate the proposed models of Akashons and Photon.

In the above diagram bold spot gives the location of spheres of tranquillities of interacting particles. Here the separating force is supplied by the disproportionately populated regions of perturbation of interacting particles unlike the Feymnman diagrams where force is supplied by unrealistic virtual particles. The study of interactions according to QED is much complicated because nothing is known there about the internal structures of interacting particles. The knowledge of internal structures have made interaction studies very straight forward.

CASIMIR EFFECT:

The matter surface is composed of very large number of positive and negative charged Akashons. In case of metal surface loose electrons are present in the surface. The electrons are associated with permanent electromagnetons which behave like particles of energy. It has already been discussed that charged particles associated with electromagnetons and two electromagnetons combine to form photon. Photon is energy packet of electromagnetic energy and hence it is concluded that source of electromagnetic energy as discussed in Casimir effect is not vacuum but electromagnetons present in the surface of the plates. The electromagnetic energy density of photon or electromagneton as already discussed is given below:

Energy density = ρ_E = dE/dV = $- hc/16\pi^2 r^4$ = $- (1.258/r^4) \times 10^{-27}$ J.m^{-3}

r is the distance form the sphere of tranquillity or as in approximation it is the distance from the surface of the plate. The energy density has got dimensions of pressure.

Pressure = ρ_E = $(1.258/r^4) \times 10^{-27}$ N.m^{-2}

The pressure falls rapidly with increase of distance form the plate's surface. The pressure formula can be put as given below:

P = $\pi hc \approx \pi hc/16\pi^3 r^4 \approx \pi hc/496 r^4$

The pressure calculated by using very complicated mathematical calculation according to Casimir effect $P = \pi hc/480r^4$. The small discrepancies in the pressure formula calculated by very simple calculation by me and complicated method used to calculate pressure is due to the fact that I have calculated pressure only for single Akashon, but so many Akashon equivalent photons are present in the surface.

The spheres of tranquillity of electromagnetons are spheres therefore pressure is directed in all directions along the lines starting from center and cutting the surface of the sphere. The radial lines can be called as lines of pressure. All the lines will not be perpendicular to the surface of plate. Most of the lines will form angle less than 90^0. If the line of pressure leaving the one surface is parallel to the line of pressure associated with the other plate, the pressure is re-inforced, therefore, magnitude of pressure exerted depends on the number of parallel lines. If the plates are parallel, only the lines which are perpendicular to the respective surfaces will be parallel. The pressure effect of these lines will be maximum, because distance r is minimum. The other lines leaving the surfaces of the respective plates at other angles can be also parallel but the pressure effect is less due to –

(1) distance of interaction of oblique lines will be more
(2) the component of force along direction perpendicular to the surface will be also less

From above, it is concluded that surfaces of the plate should be smooth for maximum number of electromagneton to be aligned together and hence maximum number of lines leaving the surface will be in perpendicular direction. On the other hand large number of cavities in the surface will lead to decreased Casimir effect. In the usual explanation, presence of two contradictory properties, smoothness and cavities in the surface are required.

From the above discussion it is concluded that Casimir effect is not vacuum property but it is the property associated with the surface.

The same energy density formula can be derived for the electromagneton energy associated with charged Akashons. The dimensions of the energy density are that of (force/area). In other words, energy density represents pressure.

It is observed that if two conducting or dialectic plates with very smooth surface are brought very near to each other attractive forces observed between the plates, and this is called 'Casimir effect'. The Casimir effect is said to be due to quantum vacuum fluctuations of the electromagnetic field between the plates. According to quantum theory, vacuum contains virtual particles which are in continuous state of fluctuations. The unbalanced force is manifested as Casimir effect. Based upon above assumptions, pressure exerted on each plate $P = \pi hc/480r^4$, here r is the distance between the plates. So, according to above line of treatment Casimir effect is a property associated with vacuum, no part is played by the surface of the plates except providing cavities. The above formula is derived by using very complicated mathematical calculations and also using unrealistic approximations.

According to me Casimir effect is not vacuum property, it is a property associated with the surface of the plates in the same manner as surface tension is the property of the surface. According to energy density formula as derived above, energy density has got pressure units. Therefore, the cause of Casimir effect is intrinsic energy (mass energy) or cosmic force which is associated with all the Akashons present in the surface. Therefore, pressure

$$\rho_E = hc/16\pi^2 r^4 = \pi(hc/16\pi^3 r^4) = \pi hc / 496 r^4$$

The small difference in the pressure calculated may be due to the fact that the above value is calculated by focusing attention on the single Akashon only.

This pressure formula has been derived very easily. The conclusion is that the Casimir effect is not vacuum property but it is the surface property. It is so because in the surface of the plates, electrical potential energy is packed as electromagneton energy. When two plates are brought close together the interaction between the intrinsic energy present in the two plates unbalances the force. Therefore, each plate appears to be exerting pressure on the other plate. The Casimir effect force is the same as strong nuclear force where electromagneton and mattenergon energies interaction provides the attractive force. In case of nucleons the distance can be very much decreased therefore, the intensity of force being distance dependent becomes very large and that is called strong nuclear force. The cause of this strong nuclear force is the same i.e. interaction between electromagnetons.

The pressure exerted by photon: It has been seen that the energy density of photon is equal to $hc/16\pi^2 r^4$. Therefore, each photon can exert pressure on other objects due to energy density and this has been observed in many phenomena taking place in space.

MAGNETIC FIELD:

As in case of Akashon special inertial temporary mattenergon is created during inertial time interval while it is moved from rest. Similarly with the moving charge packet, special magneticinertial temporary magnetoelectromagneton is created. The magnetoelectromagneton has its own sphere of tranquillity. The inertial temporary magnetoelectromagneton is the casue of magnetic field associated with the moving charge. The magnetic inertial magnetoelectromagneton remains associated for the time interval equal to inertial time even after the charge packet comes to stop. Therefore, magnetic field can be experienced for a time equal to the inertial time after the charge packet stops moving. The change in the spin of electron after the emission of photon is discussed below using the concept of electromagneton. If no magnetoelectromagneton is associated, the emission of photon from electron is given below along with the respective spins.

1. Electron $\longrightarrow$ electron + photon
   +1/2   +1/2   −1

2. Electron $\longrightarrow$ electron + photon
   +1/2   −1/2   +1

The total spin after the emission of photon in first case is − ½, which means that though the numerical value of the spin is same, there is change in the sign. In the second case there is no change in the spin after the emission of the photon, but the spin of the electron is reversed after the emission, but the reversal of the spin requires energy. The discrepancy in the electron spin sign requires that while calculating the spin change, the contribution of so far unknown spin associated factor has to be accounted for. The ignored spin factor is the contribution of the spin of inertial electromagneton. The spin of inertial magnetoelectromagneton is ± 1. If the spin contribution of the inertial magnetoelectromagneton is considered while calculating the spin of electron, there is no reversal of the spin of the electron or the total spin because, with the emission of photon, the inertial magnetoelectromagneton with opposite spin to that of the photon is also formed which cancels the change in the spin.

$$\text{Electron} \longrightarrow \text{electron} + \text{inertial electromagneton} + \text{photon}$$
$$+1/2 \qquad\qquad +1/2 \quad \pm 1 \qquad\qquad 1 \qquad\qquad \mp$$

The inertial magnetoelectromagneton is composed of the displaced shells arranged around the sphere of tranquillity in the same manner as in case of Akashon or Prakashon but with the difference that the shells of the inertial magnetoelectromagneton are much more mobile as compared to the shells of the Akashon. The shells are displaced inwards towards the center. The shells of the inertial magnetoelectromagneton are not equally displaced inwards in all the directions as in case of Akashon. The shells of the inertial magnetoelectromagneton are squeezed inwards with maximum force acting along the line perpendicular to the direction of motion of the charge packet.

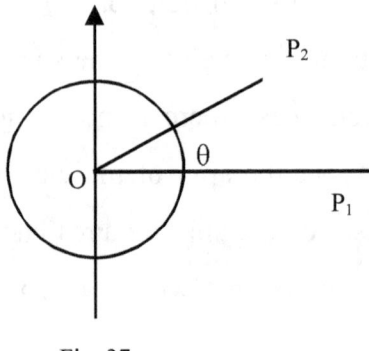

Fig. 37

The two shells P_1 and P_2 are located at the same distance from the center and the θ is the angular separation. The point P_1 lying on the line perpendicular to the direction of motion is pushed inwards with maximum force while the shell P_2 though equally placed is pushed inwards with less force. If the shell P_1 is pushed inwards with force F, the shell P_2 is pushed inwards with force FCosθ. The inertial magnetoelectromagneton is only temporary creation and is always associated with the moving charged Akashon.

The next topic of discussion is the mutual interaction of the inertial magnetoelectromagneton associated with moving charged Akashons. When the charged Akashon is moving, the inertial magnetoelectromagneton associated with it is spinning either in the clockwise or anticlockwise direction along the axis in the direction of motion. Let us adopt the convention of the clockwise rotation of the inertial magnetoelectromagneton when the negative charge packet moving away from the eye is viewed from behind and when the positive charge packet is observed in the same manner, the direction of spin of the inertial magnetoelectromagneton is anticlockwise. Only the rest of anti Akasha, excluding the shells is rotating.

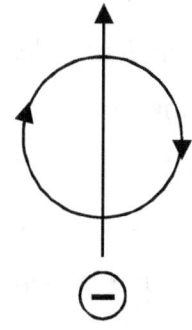

Fig. 38

Suppose two inertial magnetoelectromagneton associated with the same type of charge are moving in the same direction.

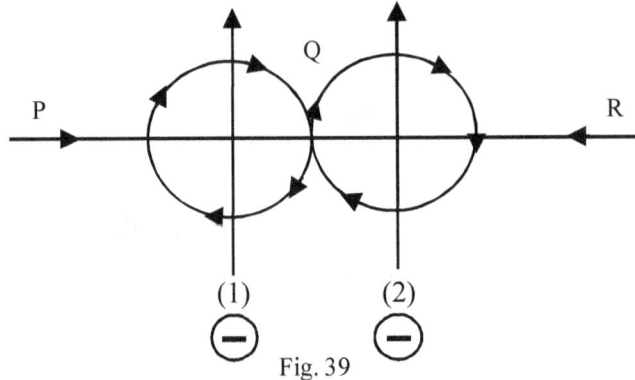

Fig. 39

At any point Q inside the parallel lines representing the direction of motion, the direction of rotational motion of the inertial magnetoelectromagneton belonging to the same type of charge is in the opposite direction while in the region Q and R, the direction of rotational motion of the inertial magnetoelectromagneton belonging to the charged Akashons (1) and (2) is in the same direction. The above observation shows that motion of anti Akasha is hampered inside the lines while it is reinforced in the regions P and R. as already pointed out, the shells of the inertial electromagnetons are mobile as compared to the shells of the Akashons which leads to the more concentration of the shells in the region of reinforced motion, therefore, the population of the shells in the region P and R becomes more as compared to the region Q. The change in the shell density leads to the unbalancing of the force acting on the inertial electromagnetons . The unbalanced force pushes the left charge packet (1) from left to right while the charge packet (2) is pushed from right to left, in other words it can be said that two similar charges moving in the same direction experience attractive force. The case of the similar charges moving in the opposite direction is illustrated as below

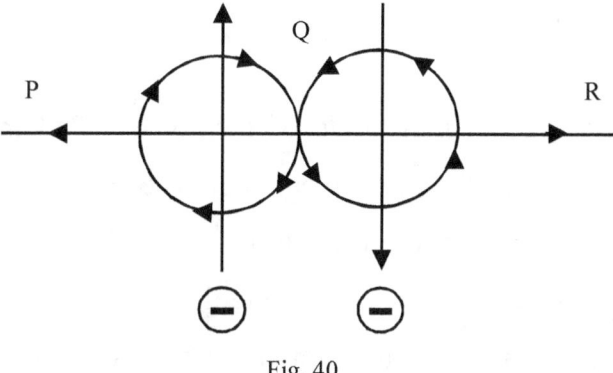

Fig. 40

Here the region Q is reinforced motion while the regions P and R are of hampered motion, therefore, the region Q has got more shell density as compared to the regions P and R, therefore, the two charged Akashons experience repulsive force. It can be easily seen that the like charges moving in the same direction experience attractive force while like charges moving in the opposite direction experience repulsive force. Next case of two unlike charges moving in the same or opposite directions is discussed.

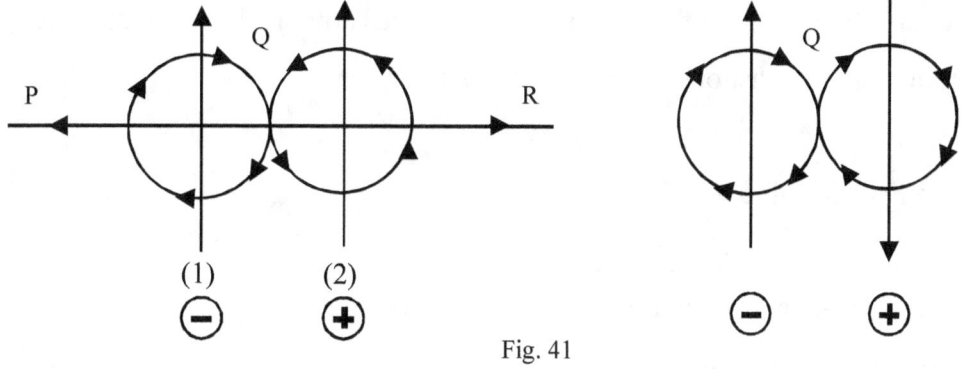

Fig. 41

It can be easily seen that the unlike charges moving in the same direction experience repulsive force while the unlike charges moving in the opposite

Evolution of Physical Laws

directions experience attractive force. The shells of Akashons or charged packets are not so mobile; therefore, spin factor is not much responsible in the manifestation of field.

Let us next find out the magnitude of the force called magnetic force experienced as a result of mutual interaction. The radius of the sphere of tranquillity of the inertial magnetoelectromagneton is the same as that of charged packet. Suppose L_1 and L_2 are the radii of the spheres of tranquillity of the inertialelectromagnetons and l_1 and l_2 are the distances of separation of the inertial magnetoelectromagneton and the geometric centers. The forces F_1 and F_2 acting on the moving charge packets are given below by the general force relation

$$F_1 = F_{02} \cdot l_1 / L_1; \quad F_2 = F_{01} \cdot l_2 / L_2$$

As already discussed, the maximum force is acting on the shells which are placed on the line perpendicular to the direction of the motion of charge packet. By adopting the same line of discussion as in case of mutual interaction of Akashons it can be seen that

$$F_{01} = F_{02} = k_E'' / 4\pi d$$
$$l_1 / l_2 = L_1 / L_2 = q_2 / q_1$$

d is the distance of separation of moving charge packet

$$l_1 = k'' \cdot q_2 \text{ and } l_2 = k'' \cdot q_1$$

Unlike the Akashon, the inertial magnetoelectromagneton has got only temporary existence because it is manifested only when the charge packet is in motion. When the charged Akashon (1) is moving and (2) is stationary and vice-versa, there is no magnetic force because no inertial magnetoelectromagneton is associated with the stationary charged Akashon. The above observation leads to the conclusion that apart from the distance of separation, k'' should be dependant on the velocity of the charged Akashon also.

$k'' = k_0'' \cdot v_1 v_2 / 4\pi d$

$L_1 = k_q / 4\pi q_1; \quad L_2 = k_q / 4\pi q_2$

The magnetic force $= F = F_1 = F_2 = (k_E'' \cdot k_0'' / 4\pi k_q)(q_1 q_2 \cdot v_1 v_2 / d^2)$

Put, $(k_E'' \cdot k_0'' / k_q) = \mu_0$, $F = (\mu_0 / 4\pi) \cdot (q_1 q_2 \cdot v_1 v_2 / d^2)$

The electrostatic force between two charges is given by the usual relation

$F(\text{electrostatic}) = (1/4\pi \epsilon_0) \cdot (q_1 q_2 / d^2)$

$1/\epsilon_0 = k_E' \cdot k_0' / k_q$

Let us next find out the relation between two constants μ_0 and ϵ_0, involved in the magnetic and electrostatic force. Putting the dimensions of the constants involved the following relations are obtained

$k_0'' = (\text{length})^2 / (\text{velocity})^2 (\text{charge})$

$k_E'' = (\text{Force})(\text{length})$

$k_E' = (\text{Force})(\text{length})$

$k_0' = (\text{length})^2 / (\text{charge})$

$k_0 / k_0' = (\text{velocity})^2$

$1/\epsilon_0 = k_E' \cdot k_0' / k_q$, $\mu_0 = k_E'' \cdot k_0'' / k_q$

Evolution of Physical Laws

on dividing, $1/(\mu_0 \in_0) = k_E'.k_0' / k_E''.k_0''$

Putting the dimensions of the constants, F for force, L for length, Q for charge and V for velocity

$1/(\mu_0 \in_0) = (F.L.L^2Q.V^2 / F.L.L^2.Q) = V^2 = $ (velocity)^2

The left hand side is universal constant, therefore, right hand side is also universal constant and can be put equal to c^2, c is the velocity of light

Therefore, $1/(\mu_0 \in_0) = c^2$

In the above case the magnetic force is calculated between two charge packets moving parallel to each other. Let us next see the magnitude of the magnetic force when the charge packets are moving in any direction.

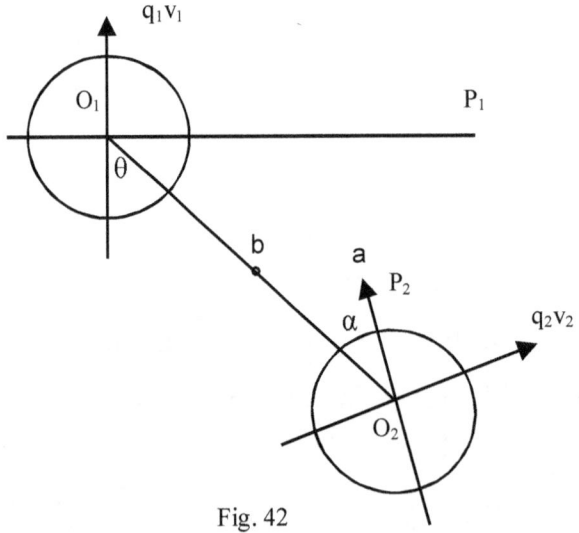

Fig. 42

As already discussed in case of inertial magnetoelectromagneton associated with the charge packet in motion, the maximum force acts on the shells located on the line passing through the center of the sphere of tranquillity and perpendicular to the direction of motion. In the above diagram, theshells located on O_1P_1 and O_2P_2 are subjected to maximum cosmic force, the situation is unlike the Akashon or charge packet shells where cosmic force acting on the shells is independent of the direction of motion. In the above diagram, the shells located on O_1P_1 and O_2P_2 are subjected to maximum cosmic force. If F_2 is the cosmic force acting on the shell located at (a) on the line O_2P_2 passing through the center of inertial magnetoelectromagneton (2) the cosmic force acting on the shell at b situated on the line O_1O_2 the force acting on shell located on O_1O_2 and at the same distance from the center as (a) is $F_2\cos\alpha$. In case of shells associated with inertial magnetoelectromagneton (1) if F_1 is the cosmic force acting on shell located on O_1P_1 the force on the shell placed at the same distance on O_1O_2 is $F_1\cos(90^0 - \theta) = F_1\sin\theta$. The force and the displacement are directly related, therefore, if the displacements of the individual shells along O_1P_1 and O_1P_2 are α_1 and α_2, the displacement along O_1O_2 will be $\alpha_1\cos(90^0 - \theta) = \alpha_1\sin\theta$ and $\alpha_2\cos\alpha$ respectively. The total displacement of the force centre (magnetic center) caused as a result of mutual interaction of the inertial magnetoelectromagneton (1) and (2) also follows the same rule. Suppose two charge packets (1) and (2) are moving parallel to each other and l_1 and l_2 are the displacements of the inertial magnetoelectromagneton centers as a result of mutual interaction. If the charge packets are not moving parallel to each other as in the above diagram, the displacement along the line O_1O_2 of the force center of the inertial magnetoelectromagneton (1) and (2) will be $l_1\cos(90^0 - \theta) = l_1\sin\theta$ and $l_2\cos\alpha$ respectively on mutual interaction. If F_1 and F_2 are the magnetic forces acting on the two charged packets, the magnitude of the force is given by the general force relation

Evolution of Physical Laws

$F_1 = F_{02}'. l_1 \sin\theta / L_1$

$F_2 = F_{01}'. l_2 \cos\alpha / L_2$

The force factors F_{02}' and F_{01}' have got the maximum values when the charge packets are moving in parallel directions, in other words when $\theta = 90^0$ and $\alpha = 0$. The values of force factors when the charge packets are moving in the parallel directions are F_{01} and F_{02}.

$F_{02}' = F_{02}\cos\alpha,$ $F_{01}' = F_{01}\sin\theta$

As in case of parallel motion

$F_{02} = k_E''/4\pi d$ $F_{01} = k_E''/4\pi d$

$F_{02}' = (k_E''/4\pi d).\cos\alpha,$ $F_{01}' = (k_E''/4\pi d).\sin\theta$

The value of l_1 and l_2 are same as in case of parallel motion

$l_1 = k_0'' v_1 v_2 q_1,$ $l_2 = k_0'' v_1 v_2 q_2$

$L_1 = k_q/4\pi q_1,$ $L_2 = k_q/4\pi q_2$

Put the values of various factors in force relation

$F_1 = (k_E''.k_0''/4\pi k_q).q_1 q_2 v_1 v_2 \sin\theta.\cos\alpha/d^2$

$F_2 = (k_E''.k_0''/4\pi k_q).q_1 q_2 v_1 v_2 \sin\theta.\cos\alpha/d^2$

$F = F_1 = F_2$

$k_E''.k_0''/k_q = \mu_0$

$F = (\mu_0/4\pi d^2). q_1 q_2 v_1 v_2 \sin\theta.\cos\alpha$

Put, $q_1.v_1\sin\theta = B$

$q_1 v_1 = q_1.dl/dt = (q_1/dt).dl = i.dl;$ $dl/dt = v_1$

Where, i is the charge passing through the line element dl in unit time or current

$B = (\mu_0/4\pi d^2) \cdot i \cdot dl \cdot \sin\theta$

The above expression is Biot's Savart's law

While discussing the nature of any type of field it has been seen that the basic requirement for the manifestation of the primary cosmic force as the secondary force called field is the separation of the force centre and geometric centre. There is no need to introduce the unrealistic concept of the exchange of virtual particles.

DUAL NATURE OF MATTER IN STATE OF MOTION:

As it has been seen the mass packet or energy packet are composed of ether shells or anti-ether shells. There are two types of ether or anti-ether, the bonded ether and free ether. In case of bonded ether the ether is mainly present in the surface of the shells and hence during oscillations the ether shells are not source of any waves. These oscillations are called independent oscillations. The value of independent oscillations depends on the distance of shell from the center of sphere of tranquillity.

In case of free ether the free ether extends beyond the surfaces of the shells or it can be said that shells are just like holes in the free ether continuum. When shell centers of free ether oscillate, the oscillations are source of wavelets having same frequency as frequency of the free ether shell because free ether is medium for carrying the waves. The oscillations of shells in free ether are called dependent oscillations. The two types of ether, bonded and free ether are interconvertible. It has been said that when Akashon is set to move the temporary region of perturbation is created where extra energy is stored. The shells present in the temporary region of perturbation becomes source of wavelets which combine to form wave packet mattenergon which exhibit dual nature. The temporary region of perturbation is composed of hybrid ether which is formed as a result of combination of free and bonded ether. The presence of hybrid ether in the miniscule temporary region of perturbation imparts dual nature to Akashon in motion. When bonded ether shell having independent oscillations change to free ether shell which merely a hole in the stationary ether continuum disturbance or waves travel in the stationary ether continuum with certain velocity. Only the miniscule region of perturbation exhibit dual nature while the permanent region of perturbation is devoid of the dual nature.

As already discussed, the force has to be applied to move Akashon and during this stage temporary region of perturbation is created inwards, even as the external force is withdrawn, the extra temporary region of perturbation remains associated with the moving Akashon. Mattenergy is stored in the temporary region of perturbation because like the permanent region of perturbation, the stretched shells present in the temporary region of perturbation can oscillate independently like the shells of the permanent region of perturbation and during this state of oscillation, the temporary region of perturbation is an extension of the permanent region of perturbation while during the second mode of dependent oscillation each shell is source of wavelets which carry away vibratory energy of the shells; to form wave packet leaving behind the shell in the stretched or displaced state only. The genesis of the dual nature of the moving Akashon is the creation of temporary region of perturbation. The hybrid Akasha shells stored in the temporary region of perturbation of electron or photon are extruded out to spread as free ether shell for the waves to travel on when wave like nature is observed but free ether shells are agin sucked in to be changed to bonded ether shells when particle like nature is exhibited. In other words photon or moving electron carries along with it free ether shells for the waves to travel on like the caterpillar track of Army tanks. Actually moving Akashon or photon behaves like combination of two particles. In case of moving Akashon, the mass of one particle is invariant mass while the second particle is formed out of the mattenergy stored in the miniscule temporary region of perturbation. Temporary mattenergon in the particle sense is highly unstable without independent existence and its existence depends on the motion akashon. The rider particle temporary mattenergon exhibit dual nature while invariant mass particle exhibit only particle nature. In case of photon, rider particle is pure charge electromagneton which exhibit dual nature.

The velocity of matter or mattenergy waves:

Let us now discuss how shells become source of radiation. Suppose the moving Akashon is brought to rest in very very short interval of time by applying force in the direction opposite to the direction of motion. As the velocity of Akashon decreases, the mattenergy stored in the temporary region of perturbation disappears by the time the Akashon is brought to rest. Suppose dt is the time interval for which the given shell undergoes dependent oscillations to become source of wavelets after that time interval the dependent oscillations state of the shells ceases to exist. The Akashon or the associated shells are in state of decreasing velocity and each shell is also a source of wavelets which are radiated out with certain velocity. v is the velocity of Akashons or shells at any instant w.r.t a point P. The given shell is source of wavelets for time interval dt as measured by the moving Akashon clock. The distance traveled by the shell or Akashon during dt should be equal to the distance traveled by the wavelet because if this condition is fulfilled, the shell will be part of the wavelet.

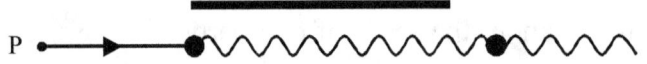

Fig. 43

Suppose P_1 is the initial position of the shell when it becomes the source of wavelets and P_2 is the position of the shell after the time interval dt, $P_1 P_2$ is the distance traveled by the mattenergy waves starting from the initial position P_1 and the next position of the shell is at P_2 due to the motion of the Akashon. P_2 is the part of mattenergy waves if $P_1 P_2$ is equal to the distance traveled by the Akashon. dt and dt_0 are the time intervals measured by the moving frame Akashon internal clock and the rest Akashon internal clock. v is the velocity of the Akashon which goes on changing because the Akashon is subjected to retardation

$$t = t_0/\sqrt{\{1-(v^2/c^2)\}}$$
$$dt = \{t_0.(v/c^2)/(1-v^2/c^2)^{3/2}\}.dv$$

If S_1 is the distance travelled by the Akashon during time interval t measured in the moving frame of the Akashon moving with uniform velocity v.

$$S_1 = v.t = v.t_0/\sqrt{\{1-(v^2/c^2)\}}$$
$$dS_1 = [t_0/\{1-(v^2/c^2)\}^{3/2}].dv$$

dS_1 is the distance traveled by the Akashon or shell in dt time interval. If V is the velocity of the mattenergy waves as measured with respect to point P. Time interval dt is measured in the moving frame of Akashon.

$$dS = V.dt$$

Evolution of Physical Laws

Putting the values of dt and dS and equating,

$dS_1 = dS$

$[t_0/\{1-(v^2/c^2)\}^{3/2}].dV = [V.t_0.(v/c^2)/\{1-(v^2/c^2)\}^{3/2}].dv$

$V = c^2/v$

When $v \to c$ the above relation proves that when Akashon is moving with velocity c, the velocity of waves, is the same as velocity of Akashon w.r.t point P. So long as v is less than c velocity of matter waves depends upon velocity of Akashon v. When $v = c$, velocity of matter waves $V = c$. Here c is constant velocity. When bonded ether shell moving with v changes into free ether shell wave disturbance having velocity $V = c^2/v$ starts from the point at which change took place. It is seen that velocity of wave disturbance depends on velocity v of shell. When $v = c$, $V = c$. For the special velocity c, the velocities of the shell and wave disturbance are equal and hence c is independent of shell or source velocity. It is only for unique velocity c that wave disturbance velocity is independent of shell velocity. The concept of maximum terminal velocity c is applicable to Akashon with zero rest mass as explained earlier, but the limit of c can be exceeded by matter waves. It is only the Akashon having finite mass that velocity cannot be greater than c. Akashon waves or matter waves starting from shellscan have velocity more than c but group velocity of wavelets is equal to Akashon velocity. It will be seen in the next topic that wavelets form wave packet which travels with velocity of light.

The major part of the Akashon in motion remains intact or particle like, it is only very tiny portion of the Akashon, the temporary region of perturbation which exhibit wave like properties.

Schrodinger equation and SHO energy:

While calculating potential energy V or U to used in calculating energy of the particle acting as SHO particle is supposed to be moving away from mean position and hence $V=1/2kx^2$. When positive value of V is put in the equation the equation can be solved for energy. The particle in oscillator is moving towards and away from the mean position and if the particle is moving towards mean position $V=-1/2Kx^2$. When negative value of V is put in the Schrodinger equation the solution is difficult or not acceptable which leads to the conclusion that Schrodinger solution is not general in nature because the motion of the oscillator is two way motionThe mattenergon quantum mechanics solution devised by me is general and needs no math.

Hydrogen atom and schrodinger equation:

The hydrogen atom is formed by the combination electron and proton initially separated by infinite distance. As electron moves towards proton electrostatic potential energy is released, this is divided between radiated energyand kinetic energy of electron. If V electrostatic potential energy. K is kinetic energy and E_m is radiated energy according to energetic $V=K+E_m$ $K=V-E_m$ V is just energy like the energy of wound up spring with positive sign.Using the above value of kinetic energy deBroglie wave equation and double derivative of progressive wave equation amplitude the the modified form of equation can be derived (laplacian operator)psi- constant (E_m-V)psi = 0, V is positive as mentioned above and E_m is the total radiated energy.When the above equation is solved the solution is the same as that of traditional equation except the value of E_m is positive. In case of traditional equation E the energy of hydrogen atom with negative sign which leads to the conclusion that hydrogen atom should be highly stable but atom is unstable and gets stability by combining with other atom. The wrong result needs modification of equation and modified Schrodinger equation has been deduced above.

DE-BROGLIE WAVE EQUATION:

As already seen when the moving Akashon exhibit wave like nature, the shells present in the temporary region of perturbation become source of wavelets and the velocity of the wavelets which start from each shell is c^2/v. If λ and ν are the wavelength and frequency of the wavelets.

$\lambda.\nu = c^2/v$

$\lambda = c^2/(v.\nu)$

It is natural to put the frequency ν of the wavelet equal to the frequency of independent oscillation of the shells from which the wavelets originate

$4\pi r \nu = c, \qquad \nu = c/4\pi r$

$\lambda = 4r\pi c / v$

$m = h / 4r\pi c, \qquad \lambda = h/mv$

It is seen that the wavelength and frequency of the wavelets depends on the value of r which is associated with the location of the shells. The wavelets originating from the shells which are placed at the same distance from the geometric center are associated with the same value of wavelength and the frequency but as already seen, the source of wavelets are the shells located at different distances in the temporary region of perturbation, therefore, the wavelets originating from different shells differs very slightly in the values of wavelength and frequency especially when the velocity is not very high because at low velocity, the temporary region of perturbation is very thin and hence location of the shells does not differ very much. The mass m depends on r but mass contribution is from the shells located just near the surface of sphere of tranquillity and hence location of shells is nearly same which leads to same mass m. The large number of wavelets which originate from the shells located in the temporary region of perturbation and hence differ slightly in the value of frequency combine together to form wave packet. The wavelets starting from the region near to the surface of tranquillity are major contributor because energy associated with these wavelets is significant. Let us next calculate the velocity of the wave packet

$$v = c \cdot \sqrt{\{1-(r^2/r_0^2)\}}$$

r is the distance of the shell from the geometric centre, there is not much difference in the value of r associated with the shells present in the temporary region of perturbation. Suppose Akashon is moving with velocity v and r is the radius of the sphere of tranquillity. r' is the location of the shell inside the temporary region of perturbation. As already expressed r' is nearly equal to r. v' is the frequency of the wavelet which originate from the shell located at distance r'

$$v' = c/4\pi r', \quad \omega = 2\pi v' = c/2r'$$
$$d\omega/dr' = -c/2r'^2$$

r and r' differ very slightly, therefore,
m = $h/4\pi rc \approx h/4\pi r'c$,
k = $2\pi/\lambda = 2\pi(m.v/h) = 2\pi.(h/4\pi r'c).(v/h) = v/2r'c$

$$v = c\sqrt{\{1-(r^2/r_0^2)\}} \approx c\sqrt{\{1-(r'^2/r_0^2)\}}$$

$$k = \sqrt{\{1-(r'^2/r^2)\}}/2r'$$

$$dk/dr = -1/[2r'^2\sqrt{\{1-(r'^2/r_0^2)\}}]$$

$$d\omega/dk = (d\omega/dr')/(dk/dr') = c\sqrt{\{1-(r'^2/r_0^2)\}} = v = \text{Group velocity}$$

The wave packet named mattenergon has got the same velocity as the velocity of the Akashon. The mattenergon has got dual nature particle like and wave like. The mattenergon can be considered to be carrier of excess energy due to motion of electron. There is nothing paradoxical about the dual nature of the Akashon in motion. The wave like nature is exhibited by the wave packet or mattenergon associated with the Akashon and the wave packet owes its existence to the creation of the temporary region of perturbation. The hybrid Akasha in the temporary region of perturbation is the source of medium for the formation of wave packet or mattenergon which is associated with dual nature. **The whole of the Akashon is not changed into wave packet as is supposed in the conventional treatment,** because if it were so the Akashon in motion should differ entirely from the Akashon at rest. The brain of the person moving with very high speed should function in peculiar manner because according to the conventional treatment the whole brain matter now exhibit dual nature. It is only the miniscule part of the Akashon, the temporary region of perturbation which exhibit dual nature. When the Akashon exhibit particle like nature, the energy of the wave packet or mattenergon is sucked in.

MICHELSON-MORLEY EXPERIMENT:

As already mentioned the velocity of photon is independent of source velocity. The ejection velocity of photon from the source is always equal to c w.r.t origin and it is unlike the relative velocity of an object which is thrown away from moving platform. The photon carries its own track (free anti-Akasha or temporary electromagneton) present in the sphere of tranquillity for the waves to propagate as the caterpillar track of an Army tank. The inherent property of light leads to negative results of Michelson Morley experiment. It is erroneously assumed that Michelson Morley experiment apparatus is floating through stationary ether. With boundary line separating the ether and apparatus and equating it with the boat floating in river.

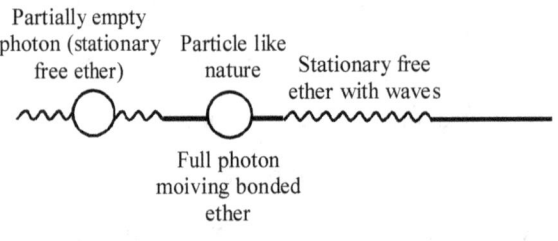

Fig. 45(a)

The negative results of the Michelson and Morley experiment set up to detect the presence of ether are there because the first principle on which the experiment is based is in itself grossly wrong. The mysterious complex nature of light has been over simplified. The wave like nature of light has been completely matched with the ripples produced in the pond of water (ether) when stone is thrown into the pond but the nature of light is not so simple. The nature of light or photon has been discussed in detail. It has already been seen in previous pages that the building material of photon is anti-ether and that of etheron is ether. When the photon exhibits particle like nature hybrid anti-ether changes to bonded anti-ether which moves with the photon, during wave nature the hybrid anti-ether changes to free anti-ether which is stationary. The waves travel in the stationary ether with velocity c relative to the point of origin of photon. The stationary ether in primordial nature is present in sphere of tranquillity as temporary electromagneton and is spread out when photon exhibits wave like nature otherwise the electromagnetic energy is carried by photon in packet form and partially empty photon has got the same velocity c. the undisturbed photon travels with constant velocity c along with the hybrid anti-ether or temporary electromagneton, it is only during interaction that hybrid ether changes form instantaneously. The photon carries the hybrid anti-ether or temporary electromagneton for the waves to travel on in the same manner as the caterpillar carries its own track. The hybrid anti-ether carried by the photon is ejected out as stationary free anti-ether or temporary electromagneton when wave like nature is exhibited. During wave nature whole of photon is not changed to stationary anti-ether it is only the hybrid anti-ether or temporary electromagneton that is converted.

The wave packet or temporary electomagneton which imparts wave nature to the photon moves with constant velocity c relative to the point of origin of photon irrespective of the velocity of the source as just proved in the previous topic. In the arrangement of Michelson-Morley experiment the photon after reflection moves within two mirrors. In conventional theory photon is considered to be different from ether but actually it is not so, ether is the progenitor of photon or Akashon. In Michelson Morley experiment it

Evolution of Physical Laws

is supposed that stationary ether is present between the two mirrors but here it is seen that free anti ether is instantaneously rolled out when wave nature is to be exhibited and waves travel with constant velocity relative to the point of origin of photon and this explains the null results of experiment without discarding the concept of ether and retaining wave nature because without waves fringes will not be formed at all.

While discussing the terminal velocity that zeros rest mass (photon) can not be used to determine relative velocity of two frames. In Michelson Morley experiment particle with zero rest mass (photon) is used to determine relative velocity of earth frame and ether frame hence negative results are expected.

The ether or Aksha is of two types **bonded** and **non-bonded**. The two states are inerconvertible. In case of non-bonded ether, ether shells are merely holes in the ether continuum or Akasha while in case of bonded ether, most of the ether or Akasha of which the shells are composed of is concentrated near the surface of the shell and its density tapers off rapidly as the distance from the shell increases. The two types of shells are used during particle like and wave like nature, bonded ether dominates during particles like nature while free ether dominates during wave like nature. The bonded ether moves with the velocity as that of Akashon but, free ether is stationary and the shells which are merely holes in the ether continuum nature are source of waves.

The negative results of Michelson-Morley experiment can be explained while retaining the concept of ether. The frequency of the oscillations of the shells associated with mass packet or energy packet plced at distance r from the center of sphere of tranquillity is given by $4\pi r v = c$. The frequency of the shell is said to be independent when the shell is tightly held and oscillations are no source of any waves. The particle nature is associated with the mass or energy under this type of oscillations. The independent oscillations take place in bonded ether. In case of oscillations in free ether each shell is source of waves having same frequency as in case of bonded ether. The waves travel with constant velocity w.r.t. surface of the free ether or anti-ether. It is through the stationary free ether that the waves which originates from the shells of free ether travel with a constant velocity w.r.t. surface. The above discussion shows that during particle like nature the particles are composed of oscillating shells of bonded ether with ether confined to the surface of shells only, while in case of wave like nature the particles are composed of free ether in which ether shells are merely holes in the ether continuum through which waves travel after originating from the shells with constant velocity relative to the stationary ether continuum surface. In case of energy packets ether is replaced with anti-ether. The bonded and free ether are interconvertible.

The hybrid ether or free ether moves with the same velocity as the velocity of photon and the waves travel with constant velocity relative to the ether surface as proved earlier independent of the source velocity. The negative results of the present experiment can be explained without discarding the concept of ether.

Evolution of Physical Laws

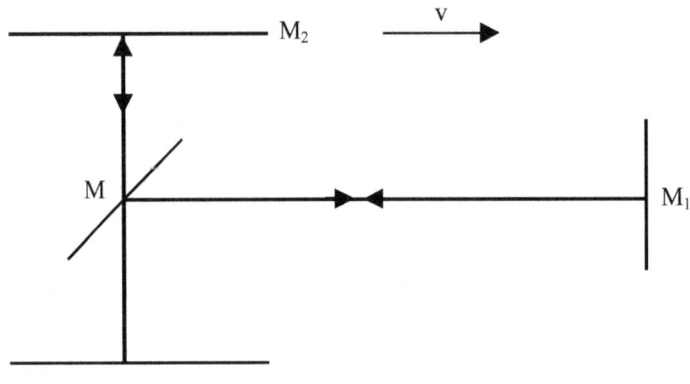

Fig. 44

In case of Michelson-Morley experiment the paths of the light traced between the mirrors after reflection are compared to the paths traveled by the swimmer while swimming upstream and downstream and also across the river from one bank to another. The river is supposed to be flowing say from left to right with velocity v. In the above simple assumptions, the velocity of the swimmer as it flows upstream is less than as it swims downstream while when it swims from one bank to the another bank the path traced is angular but this analogy of the swimmer swimming in the river is misconception when applied to the paths traveled or time taken between the reflected light between the mirrors. In case of Michelson-Morley experiment the source of light in Michelson-Morley experiment is the point situated on the surface of the mirror which is source of light. As already explained the velocity of light is independent of source velocity. The mirror surface velocity which is source of light does not affect the velovity of light as in case of sound or water waves. The production of photon or light waves as already discussed clearly proves that velocity of light is independent of the source velocity.

The above discussion leads to the conclusionthat path traced by the reflected light from M to M_2 and back from M_2 to M will be straight. Similarly, the time taken by the reflected light from M to M_1 and in the reverse direction will be same because there is no effect on the velocity of light when measured w.r.t mirror surface or velocity of light when it travels from M to M_1 or in reverse direction is not (c-v) or (c+v). The time for path traveled by the light as it moves frm M to M_2 and back will depend on the distance between mirrors.

All the above discussion proves that no fringes will be detected on the rotation of the mirror assembly. The negative results of Michelson-Morley experiment can be explained by retaining the concept of ether. The movement of ether as explained here is not like that of ether drift where only the ether near to the surface of the moving body is considered to be moving but here concept of ether is entirely different, all the parts of the apparatus like mirrors are composed of ether, the whole of bonded ether of photon moves with the velocity of light.

The simple arrangement of mirrors designed to detect the presence of ether was based on over simplified view about the nature of photon and ether, therefore, was grossly inadequate experiment. The drawbacks in the experiment led to the wrong conclusion that there is no ether, whereas in the present text all the known and unknown natural laws have been derived by using the novel concept of ether in very easy manner. Even some relations such as photon energy relation, Coulomb law of force and Newton's gravitational force law which has been assumed as such so far, without any proof have been derived very comprehensively along with many other relations. Starting with the concept of ether as discussed in the present text flawless results are got and hence the existence of ether cannot be denied. The gross blunder in Michelson-Morley experiment is that it is based on erroneous concept that there is separating boundary between the apparatus and ether whereas; according to the true nature of ether no such boundary is there because matter is manifestation of ether.

DUAL NATURE OF PHOTON:

While discussing the nature of photon it was seen that photon is doublet of pure positive and negative charge packets. Energy packets combine to form electromagneton of suitable energy. The presence of two permanenet regions of perturbation with the two energy packets is equivalent to the permanent region of perturbation of moving Akashon. As already discussed the temporary regions of perturbation are also associated with the two permanent electromagnetons. The tiny temporary regions of perturbation are extension of two permanent regions of perturbation. Tiny residual charge is used to form temporary region of perturbation. These temporary regions of perturbation are source of oscillating electric and magnetic fields or waves. By following the same arguments while discussing the dual nature of moving Akashon it can be shown that these temporary regions of perturbation are source of independent and dependant oscillations. Dependant oscillation means wave like nature and independent oscillations means particle like nature. The waves which start from region near the surface of the sphere of tranquillity differ slightly in wavelength and hence form wave packet or electromagneton in the same manner as wave packet or mattenergon is associated with moving Akashon. The exchange of energy of the Prakashon (photon) takes place through the wave packet. The Akashon or Photon left after transferring energy of temporary region of perturbation stored as mattenergon or electromagneton to waves is called partially empty akashon or photon or prakashon. When the photon exhibits particle like nature, the energy associated with partially empty Prakashon (photon) or Akashon (mass packet) play very important role during interference, diffraction and scattering etc. The partially empty Prakashon (photon) having small radius of the sphere of tranquillity and hence more energy in the permanent electromagneton form are associated with weak progressive wave packet (front) because very small amount of energy seeps through dense region of perturbation to form wave packet but in photons having large radius of sphere of tranquillity the energy carried by the wave packet and partially empty Prakashon (photon) is not so disproportionate. The tiny inertial electromagneton or hybrid ether associated with the photon imparts dual nature to photon.

As in case of moving Akashon, the whole of the Prakashon is not changed into wave packet which is born by the observation that during Compton scattering the whole of the Prakashon (photon) is not absorbed by the scattering particle.**It is only very miniscule part called the temporary region of perturbation which shows dual nature while the major part remains intact or is never a source of waves.**

Two conditions must be fulfilled for the manifestation of particle like nature of the Prakashon and the moving Akashon.

a. The presence of perturbation or mattenergon or electromagneton or progressive wave packet.
b. The partially empty photon or moving Akashon should be positioned in such a manner that the perturbation of the progressive wave packet touches the center of sphere of tranquillity in unbroken form. The center of the fully manifested Prakashon (photon) or Akashon (mass packet) is located inside the sphere of tranquillity and the chances of

the location of the center at all the points lying inside the sphere of tranquillity are equal.

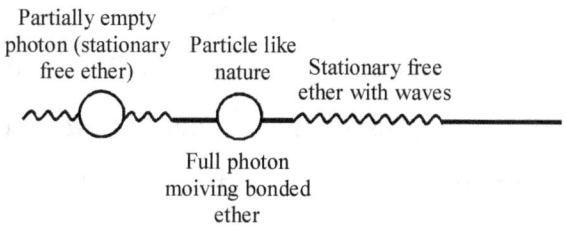

Fig. 45(a)

The progression of light can be visualized as given in the above diagram. Bonded anti-ether or temporary electromagneton moving with photon changes to stationary free ether when wave nature is exhibited it does not imply that light travels alternatively as particle or wave, in reality both the modes exist simultaneously. The method of detection of light determines the actual state in which the light exist at that time.

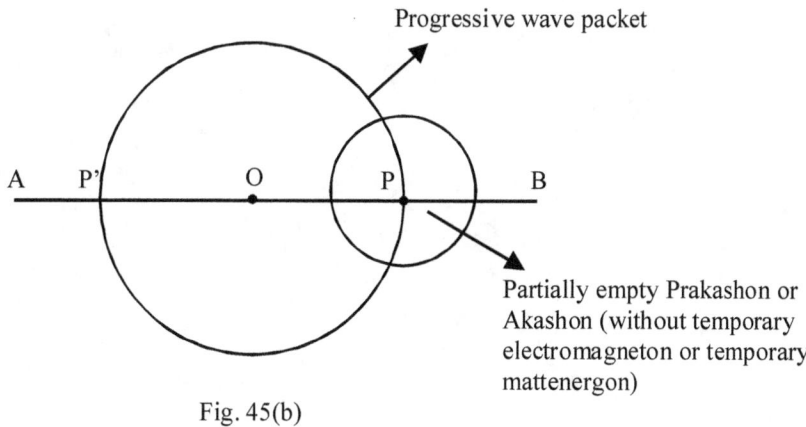

Fig. 45(b)

Suppose source O emits only one Prakashon (photon) at any instant, the progressive wave packet front is required to travel in all directions and is most intense at P, where it touches the center of the sphere of tranquillity of partially empty Prakashon (photon). The region of perturbation of the partly empty Akashon or Prakashon extends up to infinity but the oscillations of the progressive wave packet front are very intense near the point P. The energy of the wave packet front which is mostly concentrated near the point is sucked in. The photon can not be detected at point P' because of the absence of partially empty photon.

EXPLAINING THE DUAL NATURE OF PRAKASHON (PHOTON) AND THE MOVING AKASHON:

As already pointed out for the full manifestation of Prakashon or Akashon, the simultaneous presence of partially empty Prakashon or Akashon and the relatively strong perturbation of the progressive wave packet front are required. The partially empty Prakashon (photon) or moving partially empty Akashon (mass packet) and wave packet (wave like nature) always co-exist but the properties can not be manifested at the same time because for the full manifestation of the particle like nature, the wave packet is sucked in and during wave character manifestation, the wave packet is retained and the partially empty Prakashon or Akashon is debarred from full particle like manifestation.

DOUBLE SLIT DIFFRACTION PATTERN OF ELECTROMAGNETIC RADIATION AND MOVING AKASHON:

Suppose a Prakashon (photon) having radius of sphere of tranquillity equal to r is passing through slit of width d. If d is lessthan 2r or $\lambda/2\pi$ the edge of the slit will interact with the region of perturbation of the photon which results in unbalancing of the net force acting on the photon and consequently the photon is deflected from the straight path. The same thing is applicable to the motion of the Akashon. If the width of the slit is much more than 2r or $\lambda/2\pi$ the most of photons or the Akashons will pass well away from the edges of the slit without any deflection, only small fraction following the path near the edges will be deflected. Free ether (wave packet) in case of full fledged photon is without any wave train. Compared to partially empty photon with waves. The wave or wave packet or temporary electromagneton is sucked in when photon exhibit particle nature. The moving photon can be considered combination of partially empty photon and electromagneton associated with temporary region of perturbation. Electromagneton exhibits dual nature. The whole of photon or akashon does not exhibits dual nature, it is only the temporary region of perturbation which has got dual nature.

Let us discuss double slit diffraction pattern. Suppose partially empty Prakashons or Akashons are passing through the slits S_1 and S_2. The partially empty photon or Akashon can not be bifurcated. It can pass as a whole either through the slit S_1 or S_2 at any time. The height of the line perpendicular to the screen and cutting the curve drawn is measure of the partially empty Prakashon or Akashons striking the screen in given interval

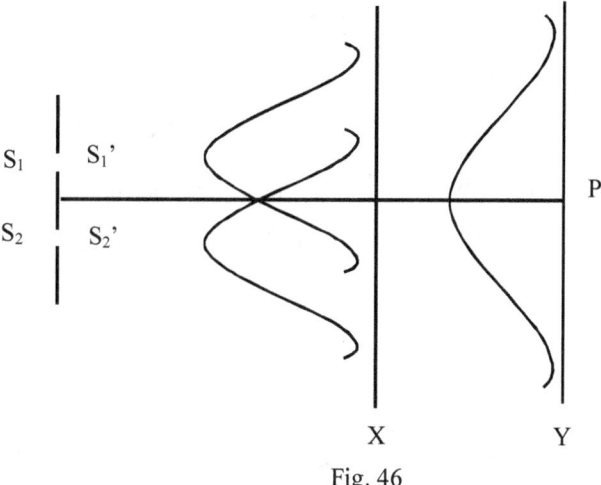

Fig. 46

The curve Y is the resultant of S_1' and S_2' associated with the single slits S_1 and S_2. As expected the maximum number of partially empty Prakashons or Akashons are striking at the point P. For the full manifestation or detection of the partially empty photon or Akashon, the presence of relatively strong perturbation or electromagneton is required.

Let us discuss the pattern traced by the wave packet alone when it passes through the slits S_1 and S_2. The wave packet can be fragmented into many wavelets unlike the partially empty Prakashon or Akashon. The wave packet associated with the partially empty Prakashon or Akashon can pass through two slits or more simultaneously at the same time after fragmentation into wavelets and this leads to phase difference in the wavelets which strike the screen after passing through the slits. The phase difference results in the region of high and low perturbation on the screen. It is the result of fragmentation of the wave packet that even a single photon passing through the slit can trace diffraction pattern. The diffraction pattern formed is not due to the passing of the single photon through the two slits simultaneously which is unrealistic. The diffraction of single photon or electron is due to the bifurcation of wave packet or electromagneton.It can be said that the wave associated with a single photon or electron interferes with itself to produce diffraction pattern.The idea of the bi-location of pohoton is absurd.

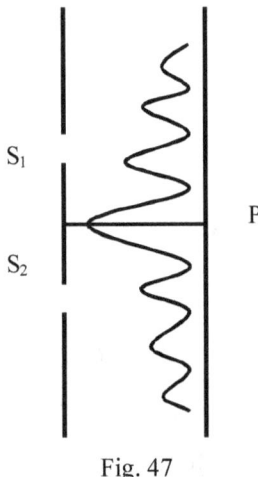

Fig. 47

The diffraction pattern produced by a single electron or photon is due to the fact that major part, the permanent region of perturbation is devoid of wave like property. It is only tiny region of perturbation which is associated with dual nature. In case of double slit diffraction using single electron and photon the partially empty photon or electron passes through only one of the slits at a time. It is the progressive wave front or electromagneton associated with the temporary region of perturbation that is bifurcated. The diffraction is produced on the screen by the simultaneous presence of partially empty electron or photon and wave packet or temporary electromagneton disturbance at a point on the screen. The bifurcated wave packets produces pattern of changing intensity on the screen for the full fledged manifestation of partially empty photon or electron. The diffraction produced by single photon or electron cannt be explained by the conventional treatment in which it is assumed that the whole of the electron or photon changes to wave packet.

The intensity of disturbance is shown in the above graph. **As already discussed, for the full-fledged detection of the Prakashon simultaneous presence of the partially empty photon and the perturbation (electromagneton) is required.** The chances of detecting the Prakashon or Akashon are very high at the point P because here both the conditions required for full manifestation are very high while at the nearby point P', the full-fledged manifestation is nearly nil because perturbation is missing. **The actual double slit diffraction pattern is got by superimposing the two graphs traced by the partially empty photons or Akashons and the wave packets (electromagneton).** The smooth curve traced by the partially empty Prakashon or Akashon becomes saw-toothed.

The same diffraction pattern is obtained if the source is so adjusted that it emits only one photon at a time, therefore, only one partially empty photon passes through either of the slits at any time. The diffraction pattern is not difficult to explain because though the partially empty photon passes through either of the slits at a time, the fragmented wave packet or temporary electromagneton can pass through both the slits simultaneously. The exposure time will have to increase so that large number of partially empty Prakashons passes through the slits. If it is possible that all the emitted partially empty photons follow the same path, no diffraction pattern will be observed in spite of the fact that regions of high and low perturbation are present. All the partially empty photons will show full manifestation at the one striking point.

If the sphere of tranquillity of the partially empty Prakashons strikes at a point of no perturbation, it can not be detected at that point but it does not imply that this particular Prakashon is lost because the sphere of tranquillity of the partially empty photon has got finite dimensions, therefore the nearby point where perturbation is present as it lies in circle of radius r drawn on the detecting screen, it will facilitate the full manifestation of Prakashon.

The particle like nature and the wave like nature are complimentary while explaining the diffraction pattern. The wave composed of wavelets traces the overall diffraction pattern but its detection or development is possible only by the participation of the partially empty Prakashon (photon). The symphony written in musical symbols of itself produces no melody because the music composed on paper is merely a guide for the musical instruments to follow. It can be concluded that for the experimental observation of diffraction or interference, both the particle like and wave like properties are required because for the spots to be marked on the screen, energy is transferred to the screen by particlesuch as photon etc.

If some photon detecting device is placed just in front or behind the slits, no diffraction pattern is produced on the screen. The graph is of the same continuous form as discussed earlier with the assumption when Prakashon (photon) exhibit only particle like nature or the bullets are supposed to be striking the screen after passing through the slits. The above special set up of the experiment leads to the conclusion that the photon now exhibit only particle like nature but it is not so, the observed results can be explained by retaining the wave like nature without active manifestation.

Suppose the Prakashon (photon) detector is placed just behind the slit when partially empty photon along with the fragmented wave packet reaches the detector, the interaction with the detector leads to the sucking in of the fragmented wave packet for the full manifestation. During the full manifestation of the photon, the wave like nature is suspended temporarily and afterwards during onwards journey to screen, the wave packet or electromagneton can be recreated but now no slits are present to divide the wave packet which is necessary condition for the production of diffraction. If the photon detector is placed just in front of the slit, the wave packet is sucked in by the partially empty photon during interaction with the detector and therefore the photon passes through the slits without the fragmentation of the wave packet because it is recreated only after the passage of the photons through the slits and hence no diffraction pattern is produced because the partially empty photon strikes the screen without fragmented wave packet.

HEISENBERG UNCERTAINTY PRINCIPLE:

The Akashon (mass packet) in motion has got dual character. So far the uncertainty principle has been discussed by focusing on the wave like properties only. It is but natural because the properties of waves has been studied in detail but ignorance about the mode of packing up of matter and energy in the Akashon and Prakashon was hurdle in focusing attention to particle like nature. In the present text for the first time internal structure has been assigned to the particles, which has made it very easy to study the uncertainty principle in straight forward manner without any philosophical implication. The discussion is also more accurate.

The Prakashon (photon) or moving Akashon exhibit dual nature due to the creation of the temporary region of perturbation. As already expressed, the shells present in the temporary region of perturbation becomes source of numerous wavelets differing slightly in the wave length which combine together to form wave packet or mattenergon which travels with the same velocity as the velocity of the Akashon.

The stretched shells present in the temporary region of perturbation transfer energy or mattenergy associated with independent mode of vibration of the shells to the wave packet or mattenergy packet while the identity of the shell is retained in the stretched state lacking independent vibrations. After the formation of the wave packet or mattenergy packet, the Akashon left behind is called partially empty Akashon.

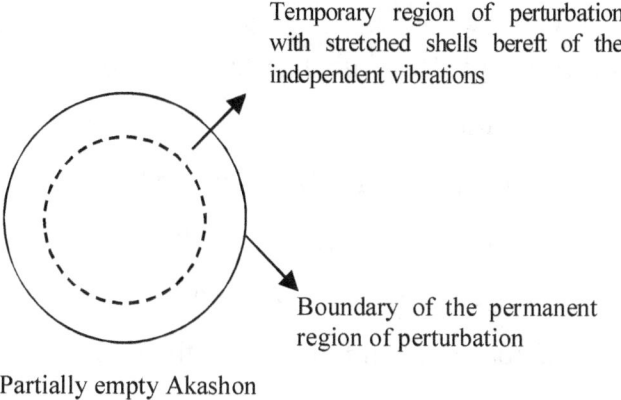

Partially empty Akashon

Fig. 48

The magnitude of many other properties like momentum, energy etc. depend on the value of radius os sphere of tranquillity, hence inherent error in the measurement of radius leads to the error while calculating these properties. There is no need to mention world simultaneously while measuring the two properties. Even measurement of one property the location of mass center which is related to the radius is not exact.

It is the temporary region of perturbation which becomes source of waves while in the conventional treatment the whole of the Akashon is supposed to be changed into wave like fuzzy state while in motion. The Akashon does not behave like point mass because mass center is free to be located anywhere inside the sphere of tranquillity or sphere of uncertainty which has got finite dimensions. It is not difficult to locate the mass center while the mass packet is at rest but the location of the mass center in case of moving mass packet is not exact because of the creation of the extra temporary region of perturbation and the major part of energy stored in the temporary region of perturbation is reached out as wave packet or mattenergy packet.

Evolution of Physical Laws

The usual process adopted to calculate the value of certain property is a formula involving variables, on which the magnitude depends, the magnitude of the property is got by using the measured values of the variables. The error in the measurement of variables lead to error in the calculation of property.The temporary region of perturbation of moving Akashon canbe compared to turbulent sea, therefore, mass center of the moving Akashon is state of flux and continuously changing position. The change in position leads to the inherent error while measuring radius of sphere of tranquillity of the Akashon.

As already discussed mass, energy momentum of the Akashon depends on only one variable i.e. radius of sphere of tranquillity. Error in the measurement of radius will lead to errors in the calculation of mass energy momentum frequency etc. because as already discussed the magnitudes of these properties is related to the radius of sphere of tranquillitybut there is natural error in the measurement of radius and hence natural errors are introduced while calculating mass energy momentum etc.The error in measurement of radius also introduces error in specifying the position of Akashon.

Natural error in measurement of radius of sphere of trqnauillity:
The radius of sphere of tranquillity of Akashson is equal to distance of mass center from the center of the shell just located on the sourface of the sphere of tranquillity. While discussing the concept of force it was seen that the distance l by which mass center is shifted away is givenby, l is also equal to

error in the position of Akashon because mass center position is position of Akashon.

$l = 4at_0^2 r_0^2/r^2 = 4at_0^2/\{1-(v^2/c^2)\} = ak_t^2/[4\pi^2 r_0^2\{1-(v^2/c^2)\}]$

$4\pi t_0 = k_t$

Since, $k_t = 1.7 \times 10^{-79}$ ms is very less therefore unless v is nearly equal to c, value of l is so small that it can be neglected, hence, for very slow velocity the error in the measurement of location is nearly zero. Even when no force is acting and hence Akashon is moving with uniform velocity, the mass center continues to oscillate with geometric center as the mean position. The extra energy created as already discussed creates temporary region of perturbation. The temporary region of perturbation is in state of flux because distribution of oscillating hybrid shells is not uniform inside the sphere of tranquillity. Hybrid shells are composed of bonded or free ether and hence can exhibit dual nature. Wavelets starting from free ether shells form wave packet which moves with the same velocity as velocity of Akashon. The radius of sphere of tranquillity of Akashon in motion cannot be exactly measured because position of mass center inside the sphere of tranquillity is continuously changing.

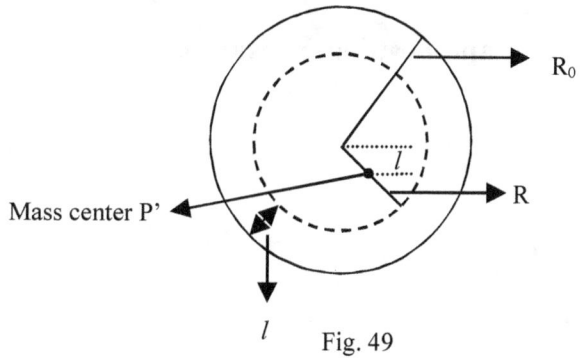

Fig. 49

Evolution of Physical Laws

The mass center of moving Akashon is shifted to point P' by distance l away from the geometric center hence, l is also equal to the error in position measurement.

Correct radius of sphere of tranquility of moving Aakashon = OP
Incorrect radius of sphere of tranquility of moving Akashon = PP'
Error – OP – PP' = *l*
Error in position measurement.

The energy in the wave packet is not equally distributed it oscillates in sequence with the oscillations of the mass center. Hence, only instantaneous location of the mass center is possible which leads to the conclusion that only instantaneous value of radius of sphre of tranquillity is possible. If as shown in the diagram P is transitory position of mass center at distance l from the geometric center. The instantaneous radius of sohere of tranquillity of moving Akashon will be (R – l), R and R_0 are the radii of spheres of tranquillity of moving Akashon and akashon at rest. l will also be error in the location of Akashon because location is pin pointed at the mass center. The unavoidable error will introduce unavoidable errors in the calculation of momentum energy etc.

The error in momentum and position measurement:
p = mv
m = h/4πRc
$v = c\sqrt{\{1 - (R^2/R_0^2)\}}$

$$p = h/4\pi R.[\sqrt{\{1-(R^2/R_0^2)\}}]$$

In the above equation to know the value of momentum only one variable R is needed to be measured. Any error in the measurement of R will introduce error in the value of momentum.

As already seen linear momentum $p = h/4\pi R[\sqrt{\{1-(R^2/R_0^2)\}}]$
The variation of p with variation in R is given by differentiating the above equation as given below. dR is the instantaneous error in the measurement of radius or position hence dp is instantaneous error in momentum.

$$dp = -(h/4\pi).[1/\{1-(R^2/R_0^2)\}^{1/2}].(dR/R^2)$$
negative sign indicates that as R increases, p decreases and vice-versa.
multiplying both sides by dR and neglecting negative sign
$$dp.dR = (h/4\pi).[1/\{1-(R^2/R_0^2)\}^{1/2}].(dR/R)^2$$

Replacing dp and dR by Δp and ΔR or Δx, where Δp and ΔR or Δx errors in the measurement of momentum and position.

$$\Delta p. \Delta x = (h/4\pi).[1/\{1-(R^2/R_0^2)\}^{1/2}].(\Delta R/R)^2$$
$$R < R_0$$

In the above expression if the velocity of the Akashon is less then the first factor is more than 1 but the second factor is very very less than 1 because ΔR is very very small for slow moving Akashon.

therefore, $\Delta p \cdot \Delta R < h/4\pi$

replacing ΔR by Δx

therefore, $\Delta p \cdot \Delta x < h/4\pi$

If velocity of the Akashon is very high nearly equal to terminal velocity $\Delta R \approx R$

If ΔR is put equal to R

The first factor is more than 1, the second factor is nearly equal to 1

$\Delta p \cdot \Delta R > h/4\pi$

The above discussion leads to two results - when velocity is much less than the terminal velocity $\Delta p \cdot \Delta x < h/4\pi$, but when velocity approaches terminal velocity $\Delta p \cdot \Delta x > h/4\pi$ for particle at rest $\Delta R = 0$, and $\Delta p \cdot \Delta x = 0$. The product of uncertainities is thus conditional.

Heisenberg's relation for energy errors of photon:

Energy of photon, $E = h \cdot c / 4\pi R$

On differentiation $dE = -(h \cdot c / 4\pi R^2) \cdot dR$

Aashon clock time

$4\pi r \nu = c$ or $4\pi R = ct$

On differentiation,

$dt = 4\pi dR/c$

$dE \cdot dt = -(h/4\pi) \cdot (dR/R)^2$

Replacing dE and dt by ΔE and Δt

$\Delta E \cdot \Delta t = (h/4\pi) \cdot (\Delta R/R)^2$, the photon moves with terminal velocity c, therefore $\Delta R = R$

therefore, $\Delta E \cdot \Delta t = h/4\pi$

Heisenberg's uncertainity principle for kinetic energy:

Kinetic energy of Aashon = $p^2/2m$

Put value of p and m, $E = hc/8\pi.[\{(1/R) - (R/R_0^2)\}]$

$dE/dR = -hc/8\pi.(1/R^2 + 1/R_0^2)$

Aakashon clock time is given by $4\pi Rt = c$

$dt/dR = -c/4\pi R^2$

Put, $dE = \Delta E$ and $dt = \Delta t$

$\Delta E.\Delta t = h/4\pi.[(c^2/8\pi).\{(dR^2/R^4) + (dR^2/R^2.R_0^2)\}]$

If the velocity of the Akashon is nearly equal to terminal velocity $dR \approx R$

$\Delta E.\Delta t = h/4\pi.[(c^2/8\pi).\{(1/R^2) + (1/R_0^2)\}]$

c is very large hence c^2/R^2 and $c^2/R_0^2 > 1$, therefore factor inside the bracket is much greater than 1, hence $\Delta E.\Delta t > h/4\pi$.

For slow velocity ΔR is very small, hence $\Delta t.\Delta E < h/4\pi r$

Uncertainty in the measurement of angular momentum and position:

Suppose mass packet of mass 'm' is moving with velocity 'v' in the circular path of radius R.

The angular momentum $L = pR$

$\Delta L = \Delta p. \Delta R$

Centripetal force = mv^2/R

The mass center is shifted away from the geometric center by distance ΔR. Therefore, due to the shift in mass center, force acts on the particle in direction opposite to the centripetal force.

Force = $F_0 \cdot \Delta R / r$

r is the radius of sphere of tranquillity of the mass packet. The above basic force relation has been discussed earlier. Equating the above forces,

$mv^2/R = F_0 \cdot \Delta R / r$
$\Delta L \cdot \Delta R' = \Delta p \cdot \Delta R'$

$\Delta R'$ is the tangential error in locating the particle along the tangent.
ΔR is the error in the measurement of position of the particle along the radius. Two errors are in directions perpendicular to each other.

$\Delta R'$ can be taken equal to ΔR
$\Delta p \cdot \Delta R = h/4\pi$, as already discussed
therefore, $\Delta L \cdot \Delta R = (h/2\pi) \cdot \Delta R$
$mv^2/R = F_0 \Delta R / R$

F_0 is the maximum force which can be applied to the mass packet.

$F_0 = hc^5 / 64\pi m^2 G^2$
$\Delta R = (64 m^3 G^2 / hc^5) \cdot (r/R)$

the super heavy Akashon N has mass $M_0 = \sqrt{(hc/4\pi G)}$

$G^2 = h^2.c^2/16\pi^2 M_0^4$

Putting the value of G^2,
$\Delta L. \Delta R = (h/4\pi)^2.(1/4\pi mc).(m/M_0)^2.(v/c)^2.(r/R)$
putting, $1/4\pi mc = r/h$ in the above equation
$\Delta L. \Delta R = (h/4\pi)(r/4\pi).(m/M_0)^2.(v/c)^2.(r/R)$

The values of squares of the fractions inside the brackets is always <1, because $m < M_0$, $v < c$, $r < 1$. The above result shows that the product of the factor associated with $h.r/4\pi < 1$. Therefore, the following result is concluded

$\Delta L. \Delta R < hr/4\pi$

The Einstein was of the opinion that the nature is not governed by chance or uncertainty and has said, "God does not play dice". The uncertainty is there, though it is not so large as projected by the conventional treatment, according to which the product of errors or uncertainties of two observable measured simultaneously is always greater than $h/4\pi$ but according to the above discussion the product of errors has upper limit $h/4\pi$ and it can be made less than this limit also. The uncertainty is there because of the wave like nature and the finite dimensions of the sphere of tranquillity.

Evolution of Physical Laws

Instantaneous error in momentum and position measurement:

$$p = (h/4\pi R).\{\sqrt{\{1 - (R^2/R_0^2)\}}\}$$

If l is the magnitude of shift of mass center away from geometric center the instantaneous value of radius of sphere of tranquillity of Akashon in motion is R – l.

Error in measurement of momentum:

The correct value of momentum,

$$p_1 = (h/4\pi R).\{\sqrt{\{1 - (R^2/R_0^2)\}}\}$$

The incorrect or instantaneous value,

$$p_2 = \{h/4\pi(R-l)\}.\sqrt{[1 - \{(R-l)^2/R_0^2\}]}$$

$$p_2 > p_1$$

Instantaneous uncertainty Δp in momentum measurement,

$$\Delta p = p_2 - p_1$$

$$= (h/4\pi)[[\{1/(R-l)\}.\sqrt{[1 - \{(R-l)^2/R_0^2\}]} - \{(1/R).\sqrt{\{1 - (R^2/R_0^2)\}}\}]$$

l is the instantaneous natural uncertainty in position measurement. Multiply both sides by l.

$$\Delta p.l = (h/4\pi)[[\{l/(R-l)\}.\sqrt{[1 - \{(R-l)^2/R_0^2\}]} - \{(l/R).\sqrt{\{1 - (R^2/R_0^2)\}}\}]$$

$$\approx (h/4\pi)[\{l/R(R-l)\} + \{(1/2)(l/R^2)\}]$$

Put the factor inside the bracket = k and Δx = l respectively.

Δp.Δx = (h/4π).k

It has been discussed earlier in force topic, magnitude of l depends on force applied or velocity. l increases with increase in velocity of Akashon. When Akashon is at rest, l = 0.

Δp.Δx = 0.

For Akashon moving with slow velocity l< R, hence k < 1

Therefore, Δp.Δx < h/4π

For Akashon with maximum terminal velocity R→ 1, hence k is large,

Δp.Δx > h/4π

For velocities less than terminal velocity but not very slow

k > 1, Δp.Δx > h/4π

For certain value of l, Δp.Δx = h/4π

It can be said that for Akashon moving with high velocity Δp.Δx > h/4π or in general Δp.Δx < h/4π or Δp.Δx ≥ h/4π or Δp.Δx = 0, which value is applicable depends on the velocity. Here the values used are not average values but values are instantaneous. The above relation shows that Δp.Δx is conditional depending on velocity. The uncettanities Δp and Δx are not inversely related. To know the exact value of momentum, exact location of mass center is needed but mass center is in state of flux, therefore, there is inherent error in the measurement of position which leads to inherent error in the measurement of radius as a consequence of that there is error in the calculated momentum. The error in Δp and Δx are very very less. The errors do not approach infinity as is some time incorrectly expressed.

Natural error in energy measurement of Photon:

Correct value of energy $E_1 = hc/4\pi R$

Incorrect value of energy $E_2 = hc/4\pi(R-l)$

As in case of momentum measurement following the same procedure,

$\Delta E = (hc/4\pi)\{l/R(R-l)\}$

$l = \Delta x$

$\Delta E . \Delta x = (hc/4\pi)\{l^2/R(R-l)\}$

In case of photon, $l \to R$,

$\Delta E . \Delta x > hc/4\pi$

The natural error in mass measurement:

Correct value of mass $m_1 = h/4\pi Rc$

Instantaneous value of mass $m_2 = h/4\pi(R-l)c$

$\Delta m = m_2 - m_1 = (h/4\pi Rc).\{l/(R-l)\}$

Multiply both sides by l,

$\Delta m . l = (h/4\pi Rc).\{l^2/(R-l)\}$ for very high velocity $R \approx l$

$\Delta m . \Delta x \geq h/4\pi cR$

Natural error in time measurement:

The oscillation of shell present in the surface of sphere of tranquillity can be used as time measuring device. The time interval between successive ticks can be taken as standard time.

$4\pi v = c$

$4\pi R/t = c$

$t_1 = 4\pi R/c$

Instantaneous time $t_1 = 4\pi R/c$ and $t_2 = 4\pi(R-l)/c$

Instantaneous error $\Delta t = t_1 - t_2 = 4\pi l/c$

Instantaneous error in energy $\Delta E = (hc/4\pi).\{1/(R-l)\}.R$

Multiply ΔE and Δt,

$\Delta E . \Delta t = (h/4\pi).\{4\pi l^2/R(R-l)\} = (h/4\pi).k$

In case of photon, $l \rightarrow R$, $k > 1$

Therefore $\Delta E.\Delta t \geq h/4\pi$

There is nothing mysterious about the uncertainties which are the results of creation of temporary region of perturbation in state of flux.

The above relations show that Δp and l or Δx are related. As l or Δx increases Δp also increases, there is no inverse relationship between Δp and Δx as is wrongly inferred. The conventional conclusion is grossly wrong. As already discussed for certain value of l, k = 1 and $\Delta p.\Delta x = h/4\pi$, but this does not lead to the conclusion that Δp and Δx are inversely related.

In the above discussion instantaneous values were used. The results can also be got by using the average values.

Average error:

For discussing uncertainity principle the previous method is more comprehensive, however alternative approach may also beconsidered. Let us see the average uncertainty in the measurement of the radius of the sphere of tranquillity of the moving Akashon. Suppose R_0 and R are the radii of the spheres of tranquillity of the Akashon at rest and the Akashon in motion respectively.

The radius of the sphere of tranquillity of the moving Akashon is the distance from the center to the inner spherical surface in the temporary region of perturbation up to which the shells having independent oscillations are present. To measure the above mentioned inner boundary, the independent oscillations of the shells present in the region of perturbation are restored by sucking in the energy associated with the wave packet but the reverse process of the inflow of the wave packet energy is not instantaneous, therefore, the inner spherical boundary up to which oscillating shells are present moves continuously inwards and hence the distance from the center to the moving boundary or in the other words the radius of the sphere of tranquillity of the moving Akashon is not the same when measured at different time intervals because temporary region of perturbation is in state of flux and is not equally populated at all times.

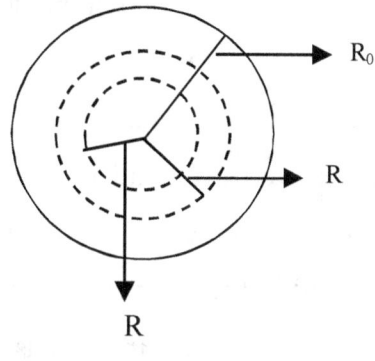

Fig. 50

R' is the transitory inner spherical boundary which changes from R_0 to R or if the observations for the measurement of the radius of the sphere of tranquillity of the moving Akashons are repeated for long time, the average radius of the moving Akashon is $(R+R_0)/2$. The radius of the moving Akashon can not be measured in one go because the return of the energy or the mattenergy associated with the wave packet is not instantaneous process. The shells with restored mode of independent vibrations are also not instantaneously distributed uniformly in the temporary region of perturbation, which leads to the shifting of the mass center away from the geometric center, but during this shifting away no force is involved because the stretched shells are uniformly distributed in the temporary region of perturbation. The shifting away of the mass center is due to unequal distribution of the shells with restored independent mode of vibrations. When the inflow of the wave packet energy starts, it first enters the right part portion of temporary region of perturbation because it is nearest to the wave packet if the Akashon is moving from left to right. The shifting away of the mass center is pronounced and directional for the moving Akashon due to creation of temporary region of perturbation where the distribution mass is subject to incessant change. The maximum shift in the mass center occurs when all the energy of the wave is transferred to the temporary region of perturbation because at that stage there is maximum unequal distribution of shells with restored independent vibrations. The minimum shift in the mass center is zero when the sucking in process of the energy is just initiated, therefore, the shift in the mass center varies from 0 to L', here L' is the maximum shift. The average shift in the mass center or the location of the moving Akashon is L'/2 = L.

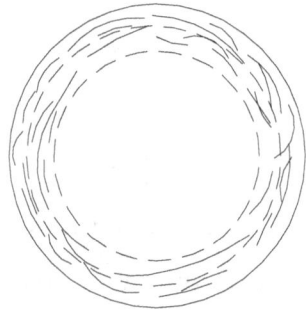

Fig. 51

The diagram shows unequal distribution of oscillating shells.

Let us see next how the momentum of the Akashon can be determined when the minimum number of measurements are required. In usual process to know the magnitude of momentum, three variables i.e. mass, distance and time measurements are required.

Velocity of the Akashon = $v = c \cdot \sqrt{\{1 - (R^2/R_0^2)\}}$

Mass of the Akashon = $m = h/4\pi Rc$

R is the radius of the sphere of tranquillity of the moving Akashon

Momentum $= p = mv = h/4\pi R \sqrt{\{1 - (R^2/R_0^2)\}}$

In the above relation, to know momentum, measurement of only one variable i.e. radius of sphere of tanquillity of moving Akashon is needed if radius of the sphere of tranquillity of Akashon at rest is known.

In the above relation for the calculation of the momentum, the measurement of only R is required provided R_0 is known. We have seen earlier that the radius of sphere of tranquillity of the moving Akashon is in state of flux, the average value is $(R+R_0)/2$ while the exact value is R.

The exact momentum $= p_1 = h/4\pi R \sqrt{\{1 - (R^2/R_0^2)\}}$

The average value of momentum,

$p_2 = h/[\{4\pi(R+R_0)/2\} \sqrt{1 - \{(R+R_0)^2/4R_0^2\}}]$

$R+R_0 > 2R$, therefore, it can be easily seen that $p_1 > p_2$

The error or uncertainty in the measurement of momentum

$\Delta p = p_1 - p_2$

The average displacement of the mass center away from the geometric center is L, therefore, average error or uncertainty in the measurement of the moving Akashon is $L = R + (R_0/2)$, because minimum and maximum values of L lies between R and R_0.

$\Delta p = p_1 - p_2$

$\Delta p = h/4\pi[1/R\sqrt{\{1-(R^2/R_0^2)\}} - \{2/(R+R_0)\}\{\sqrt{1-(R+R_0)^2/4R_0^2}\}]$

$\Delta p.L = h/4\pi.(L/R)[\{\sqrt{1-(R^2/R_0^2)}\} - (2R/R+R_0).\sqrt{\{1-(R+R_0)^2/4R_0^2\}}]$

L<R, because mass center can not be located outside the sphere of tranquillity. The general relation proves that $\Delta p.L$ or $\Delta p.\Delta x$ are not always inversely related. The product $\Delta p.\Delta x$ depends on velocity.

For slow velocities, $R = R_0$

L/R <1, the factor inside the bracket is also less than one. Therefore R.H.S. of the above relation is less than one.

Put, $L = \Delta x$

$\Delta p \Delta x \leq h/4\pi$

Δp and Δx or L has been found out separately, though from the error equation it is seen that if L is small Δp will be more and uncertainty in position $L = \Delta x$ will be less, but there is no direct proportional relation between the two errors. The result is opposite to the conventional relation which states $\Delta p\Delta x > h/4\pi$. It implies that the product of two errors is never zero and it also leads to ridiculous conclusion that when the particle is at rest, therefore $\Delta p = 0$, the error Δx in position is infinite, but the relation derived above shows that when the particle is at rest $\Delta x = 0$ because in stationary Akashon there is no shift in the mass center, therefore, $\Delta p. \Delta x$ is equal tozero.

Evolution of Physical Laws

The conventional relation is arrived at by supposing that the whole Akashon (mass packet) behaves as wave packet while in motion but the result deduced in the present treatment is based on the observation that it is only the temporary region of perturbation created as a result of motion which exhibits wave like character and hence small wave packet while the rest of the Akashon remains as such. The inverse relation between Δx and Δp is not assumed apriori. The assumption that the whole of the Akashon while in motion is changed into wave packet would lead to the consequences which are not actually observed. If it were so that the whole of the matter would change into wave packet while in motion, the brains of the astronauts moving at very high speed should start functioning in very peculiar manner.

The conventional treatment disregards the effect of velocity on the product of uncertainties or errors, the product is always greater than $h/4\pi$ which does not appeal to the mind, but in the treatment presented above the product is velocity dependent because the product relation involves R which is related to velocity. When the Akashon is moving with very low velocity, R is nearly equal to R_0 and L, the shifting away of the mass center is nearly equal to zero.

$R \approx R_0$
$L \approx 0$

Putting the above values in the product relation it can be seen
$\Delta p \Delta x = 0$

For very high velocity $R \ll R_0$

Hence, first factor is much greater than 1, though factor inside the bracket is less than 1, the product is more than 1, hence $\Delta p.\Delta x > h/4\pi$

$L = R = \Delta x$

$R = R_0 \sqrt{\{1 - (v_c^2/c^2)\}}$

v_c being terminal velocity,

$v_c \approx c$ for microscopic particles,

therefore, R is nearly equal to zero

L/R is large

When the above values are put in the product relation the result shows that $\Delta p \Delta x > h/4\pi$

It is concluded that unlike the conventional result, the product of errors varies between 0 to greater than $h/4\pi$. The product of errors relation has been studied by applying it to particles moving with high speed (terminal velocity) and hence the results agree with the relation

$\Delta p \Delta x > h/4\pi$

But for slow moving particles, the result should agree with $\Delta p \Delta x < h/4\pi$. The result that $\Delta p \Delta x \geq h/4\pi$ is derived by supposing that whole of the Akashon exhibits wave like nature and this result is erroneous, only temporary region of perturbation is associated with wave like property hence magnitude of errors is less.

The present treatment states that the product of errors can be made less by slowing down the motion of the Akashon.

Errors in the measurement of energy associated with the moving Akashon can also be deduced. The extra mass associated with the moving Akashon is given below

The extra mass = $h/4\pi c \, (1/R - 1/R_0)$

The extra energy = $\Delta mc^2 = hc/4\pi(1/R - 1/R_0)$

E_1 = the correct energy = $hc/4\pi(1/R - 1/R_0)$

E_2 = the incorrect energy = $hc/4\pi\{2/(R+R_0) - 1/R_0\}$

The unavoidable error

$\Delta E = hc/4\pi R\{(R_0 - R)/(R_0 + R)\}$

Error in the measurement of time:

Now suppose two signals with the speed of light are sent from the mass center and geometric center towards a point fixed on the line along the direction of motion. The time interval between the signals sent from the displaced mass center and un-displaced geometric center is L/c, because L is the average unavoidable displacement error of the mass center. The error in the measurement of radius of sphere of tranquillity is associated with the error in the measurement of time. Light signal is used as time measurement device. Δt is the error in the measurement of time.

Evolution of Physical Laws

$\Delta t = L/c$

$\Delta E . \Delta t = h/4\pi . (L/R)\{(R_0 - R)/(R_0+R)\}$

L/R < 1, for slow velocity and the factor inside the bracket is also less than one

$\Delta E . \Delta t < h/4\pi$

If the velocity is very high as already seen

R is very very small, L/R is large

$\Delta E . \Delta t > h/4\pi$

MATTENERGON QUANTUM MECHANICS:

Mattenergon quantum mechanics reveals causative explanation for quantum condition. In mattenergon quantum mechanics stress is laid on the behaviour of mattenergon associated with moving Akashon but in wave quantum mechanics the focus is on waves associated with the moving Akashon. If mattenergon quantum mechanics is used energy problems can be solved by using classical mechanics and thus dispensing with the complicated mathematics used in wave quantum mechanics for solving the same problems. In mattenergon quantum mechanics Schrodinger Wave equation is not required to solve the energy problems because problems are solved by focusing on the mattenergon and thus disregarding wave nature. The mattenergon quantum mechanics can give the real physical explanation of the results contrary to the wave quantum mechanics which relies on complex mathematics and gives bizarre interpretation of the results. The interpretation is not real in nature it is only based on mathematical jugglery. In classical mechanics the mass of the particle is supposed to be located at a point, and therefore, the future position of the mass center can be predicted with absolute certainty but the above treatment or classical mechanics has got only limited scope. Einstein was strong supporter of the second line of thinking. To discuss the motion of the Akashon (mass packet) special treatment is required because in case of the Akashon or the mass packet in motion, the sphere of tranquillity where the mass center can be located has got finite dimensions and there is no absolute certainty where the mass center within the sphere of tranquillity can be located. The uncertainty in locating the mass center leads to the idea of probability of locating the mass center. The above concept about the uncertainty is bizarre according to classical concept but it is not so mysterious because the concept of uncertainty automatically rushes to the mind when the mass distribution in the region of perturbation around the sphere of tranquillity and the mattenergy waves associated with the moving Akashon arein state of flux. The concept of probability while locating the mass center is significant only in case of Akashon having very very small mass because the mass and the volume of the sphere of tranquillity are inversely related. In case of heavy Akashon, the sphere of tranquillity is nearly a point, therefore, concept of probability is meaningless.

It has already been discussed while discussing the nature of force that the effect of impressed force to move the Akashon (etheron) as a whole is not instantaneous. During inertial time interval, the mass centre and the geometric centre are separated while the Akashon as a whole does not move. The potential energy generated as a result of stretching away of the mass centre is stored up as inertial temporary mattenergon. We also know that a permanent mattenergon is associated with the Akashon and on uniform motion a temporary mattenergon is added to the internal mattenergon. The potential energy stored up in the inertial mattenergon is changed into kinetic energy of the Akashon when it starts moving after the time interval.The temporary mattenergon associated with the temporary region of perturbation disappears as soon as the Akashons comes to rest but the life of inertial temporary mattenergon extends a little further till the inertial time interval is over. The creation of motion mattenergon starts near the surface of the sphere of tranquillity diametrically opposite to that point located on the surface of sphere of tranquillity where force is applied. The creation of inertial temporary mattenergon opposite to the point of impressed force is preferred because here the shells are less as compared to region on the opposite direction. The energy or mass centre of the inertial mattenergon is shifted in a direction opposite to the direction in which the mass of Akashon is shifted. The energy associated with mattenergon can be called mattenergy. During initial stages of creation of mattenergon the mattenergy is not uniformly distributed around the sphere of tranquillity of inertial mattenergon, therefore energy centre is shifted away. After the full growth of mattenergon the energy centre which was previously in stretched state as long as akashon is subjected to force starts oscillating in the same manner as stretched rubber band does on releasing the stretching force hence the akashon moving with uniform speed after removing the force is source of waves. The inertial mattenergon is the source of wavelets. The mattenergon associated with moving Akashon is made up of inertial mattenergons (wave packets). The wavelength of waves associated with the inertial mattenergon is the same as the wavelength associated with the moving Akashon. The velocity of the waves also equals to the group velocity of wavelets starting from the temporary region of perturbation. The group velocity is equal to the velocity of the Akashon, v, ν is the frequency of the radiation associated with the mattenergon. The wave packet and mattenergon are interconvertible. λ is the wavelength of wave packet and mattenergon center is source of waves having velocity v frequency ν and wavelength λ.

Evolution of Physical Laws

The Calculation of Kinetic Energy of the Akashon:

$\nu\lambda = v$

$\lambda = h / mv$

$\nu = mv^2 / h$

Kinetic energy of the Akashon $= (1/2).mv^2 = (1/2).(h\nu)$

The relation is similar to Planck's energy relation for photon.

Kinetic energy can also be calculated of pure mass by using the concept of self gravitationalpotential energy stored up as mattenergon. Potential energy stored up has already been calculated. As already seen self electro static potential energy of pure electromagneton is equal to self gravitational potential energy of mattenergon and value is given below

$E = hc/8\pi r^2$

The radius of the sphere of tranquillity of matenergon is r_0.

Potential energy stored up in mattenergon is equal to the kinetic energy of Akashon.

If ν is the frequency of oscillation, $r_0 = c/4\pi\nu$

Put the value of r_0

E = Mattenergon energy = Kinetic energy = ½ hν

The above equation shows that to find the kinetic energy of the Akashon only ν needs to be measured.

When the force acting on the Akashon is withdrawn, the Akashon starts moving with uniform velocity. At this stage all the three centres – mass centre, geometric centre, and mattenergon centre overlap.

As already discussed the numerous oscillating shells are source of wavelets. The wavelets are progressive simple harmonic waves differing slightly in wave length and frequency, as a result of which the wavelets combine together to form wave packet. It is thus seen that the mattenergon or wave packet are associated with the moving Akashon. The inertial mattenergon is created during inertial time interval when the mass center changes position by action of force. The wavelets are grouped together by the De-Broglie wave equation

$$\lambda = h/mv$$

By using the above equation and the wave equation, the time independent equation called Schrödinger wave equation can be derived.

$$\nabla^2 + (8\pi^2 m/h^2).(E - v)\psi = 0$$

Ψ is the amplitude of the mattenergy waves. Mattenergon has got dual character.

The problems related to the energy of the Akashon under different conditions can be solved in two ways, **either by focusing the attention on wave packet or mattenergon or particle nature.**

Evolution of Physical Laws

If the wave packet is used to discuss the energy problems, the Schrödinger wave equation is the useful tool. The Schrödinger wave equation involves ψ which can be taken as equal to the amplitude of the waves. The waves can be named as matter or mattenergy waves.

To detect the Akashon (mass packet) in motion, the following two conditions must be satisfied simultaneously –

a. The presence of mattenergy wave or perturbation associated with the wave packet or mattenergon.
b. The presence of partially empty Akashon.

The Akashon is fully manifested when the wave packet is sucked in by the partially empty Akashon.

It can be easily seen that ψ is measure of the above two conditions. As already discussed due to the finite dimensions of the sphere of uncertainty there is uncertainty in locating the mass center, therefore ψ is associated with the probability of locating the mass center or the Akashon. When the two events are associated simultaneously for the particular event to happen, the total probability is the product of the probability associated with the two separate events. The ψ is measure of probability for each of the above mentioned two events to happen, therefore, the total probability of locating the Akashon is $\psi \times \psi = \psi^2$.

Actually there is not much uncertainty in locating the Akashon because according to the conventional treatment, the whole of the Akashon (mass packet) in motion is supposed to be changed into wave packet but this notion has been discarded in the treatment presented here, because as already discussed it is very miniscule part, the temporary region of perturbation of the Akashon in motion that exhibit wave like nature, while the rest of the Akashon remains intact and exhibits particle like nature. In case of very very light particles the radius of the sphere of tranquillity is very very large. When the mass approaches zero, radius of sphere of tranquillity approaches $+\infty$ or $-\infty$. If the uncertainty is discussed in general, the region of uncertainty extends from $+\infty$ to $-\infty$. God does play dice but with restraint. In case of particle with finite mass mass center is located within the sphere of tranquillity having very very small volume.

The energy problems of the Akashon can be calculated either focusing on the wave like nature (Schrödinger wave equation) or considering the particle like nature. In the present text the main focus is on the particle like nature while there is only passing reference to the wave nature. **The same energy problem can be solved very easily if particle like nature is considered while by using Schrödinger wave equation, very complicated mathematics has to be used to arrive at the same result.** The method adopted is very simple and the veil about the mystery or strange behavior of the Akashon is lifted.

It is said that classical mechanics is special branch of quantum mechanics when quantum number has got large values, but here it will be shown that all the **energy problems of microscopic particles can be solved by using classical mechanics.** Very complicated mathematics is required to solve the same problems if quantum mechanics is used. Microscopic state of matter is not different from macroscopic state except that that extent of subdivision has gone too far. The laws of classical mechanics should be applicable to deduce energy problems at microscopic and macroscopic level. It will be proved in the following pages that classical mechanics is successful and more elegant in solving energy problems of elementary particles. There is no need to discard common sense because there is physical interpretation of the result got.

As already discussed the mattenergon is created during inertial time interval and just at the time of creation, the energy center of the mattenergon is shifted away from geometric center in the direction opposite to the direction of the drifting away of the mass center under the influence of force.

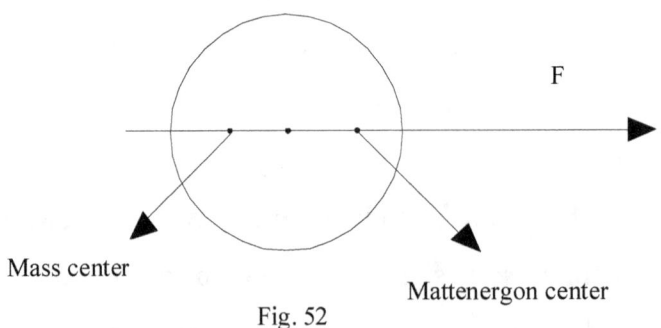

Fig. 52

The energy center of the mattenergon is shifted away at the time of the birth because uniform distribution of energy is not instantaneous process. The energy (mass) created as a result of motion is associated with the wave packet or mattenergon. If the force acting on the Akashon is withdrawn after some time interval, the Akashon starts moving with uniform velocity the mattenergon center, mass center and geometric center overlap after the withdrawl of force. The oscillating mattenergon center is source of mattenergy waves which impart dual nature to Akashon. When the Akashon is moving with uniform velocity, the mattenergon energy is not radiated out, it remains associated with the mattenergon or Akashon.

When the Akashon is moving with uniform speed, theinertial mattenergon centre is set into oscillations because under the influence of force the mattenergy center was stretched away like an elastic string and the withdrawal of force leads to state of oscillations. The inertial mattenergon center can be taken to be source of waves with wave length λ and frequency ν so that,

$$\lambda.\nu = v$$

As already seen, kinetic energy of the Akashon is ½(hν)

To know the value of ν under different conditions, the restraint is imposed on the motion of the Akashon and the restraint imposed is quite natural and appeals to the mind. The concept of mattenergon is used to calculate energy, hence the new approach is named as mattenergon quantum mechanics.

The restraint expresses that the state of oscillating center of theinertial temporary mattenergon remains unchanged when the Akashon returns to the initial state after being subjected to the various changes.

Evolution of Physical Laws

If the above restraint is not obeyed i.e. there is change in the mattenergon after the Akashon reaches the starting point after following different paths, the first law of thermodynamics will be violated because it is the state of mattenergon which is associated with the energy of the Akashon. The mattenergon is the store house of extra energy.

Let us next calculate the value of energy of Akashon under different conditions. Classical mechanics will be used to calculate energy under different conditions.

The paradigm that classical mechanics cannot be used to get the same results as derived by the application of Schrodinger wave equation is demolished.

PARTICLE IN A BOX:

The Akashon is moving in a box with impervious walls or in term of energy, the potential energy inside the box is zero, therefore, it can move freely while just outside the walls of the box potential energy is infinite, therefore, the Akashon or the mass can not penetrate the walls. When the Akashon reaches the wall of the container, it is reflected back. To simplify the problem let us suppose that the motion of the Akashon is confined along one direction only, therefore, the Akashon is moving back and forth on being reflected at the two opposite faces. Suppose a is the distance between two opposite faces. The Akashon starting from one face with velocity v and being reflected back at the opposite face just reaches starting point after traveling distance equal to 2a. Since the Akashon reaches the starting point, the oscillating center of theinertial mattenergon should be in the same state as it was when it just started moving towards the opposite face, theinertial mattenergon center should make complete number of oscillations during the time required by the Akashon to travel distance 2a. Suppose number of oscillations completed during this time interval is n where n = 1, 2, 3, 4,…... The time required to travel 2a distance is 2a/v, therefore, during 2a/v time interval n oscillations are completed.

The frequency $\nu = nv/2a$, $\lambda = h/mv$

Put value of $v = h/(m\lambda)$, $\nu = nh / 2am\lambda$, $\nu\lambda = v$
Put $\lambda = v / \nu$, $\nu = nh\nu / 2amv$,   put $v = nv/2a$
$\nu = n^2h/4a^2m$
$E = \frac{1}{2}(h\nu) = n^2h^2/8ma^2$

Evolution of Physical Laws

If the potential energy of the barrier is greater than the energy possessed by the Akashons, all the Akashons are expected to be reflected back at the energy barrier but it is observed that some Akashons can penetrate the energy barrier though the energy carried by the Akashon is less, but if the nature of motion is considered, it can be seen that there is nothing sensational about it.

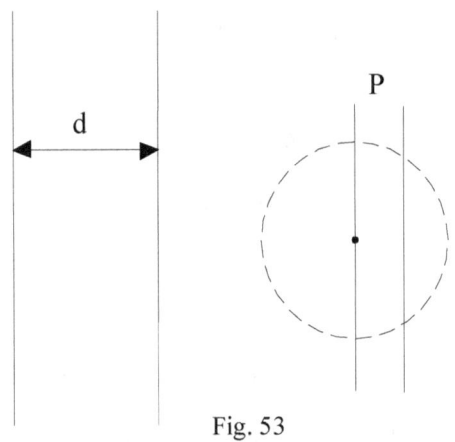

Fig. 53

Suppose d is the thickness of the energy barrier to be crossed by the Akashon. When the Akashon approaches the energy barrier, the force from right to left act on it, therefore, the mass center is pushed away from left to right and if the energy barrier is thin, the mass center can also be located in the region P or beyond the energy barrier which means that the Akashon has crossed the energy barrier instead of suffering reflection at the energy barrier. When all the three centre geometric centre, mass centre, and mattenergon centers overlap the Akashon can not be detected, because detector needs energy exchange with the Akashon and that takes place through wave amplitude, but wave amplitude is zero when the Akashon is in the above mentioned state.

SIMPLE HARMONIC OSCILLATOR:

By using the mattenergon quantum mechanics, the energy of S.H.O. can be derived without using any mathematics at all. The energy result is the same as derived by using complex mathematics in wave quantum mechanics.

When an Akashon acts like S.H.O. it executes two types of oscillations

1. **External oscillations:** The Akashon executes simple harmonic oscillations having external points fixed as extreme and mean positions. Mattenergon associated with external oscillations is also oscillating.

2. **Internal oscillations:** Inertial mattenergon is associated with the moving Alashon. Inertial mattenergon and mass centers oscillate inside the sphere of tranquillity having geometric center as the mean position. The amplitude is very less. Inertial mattenergon is associated with the internal oscillations.

Calculation of S.H.O. Energy:

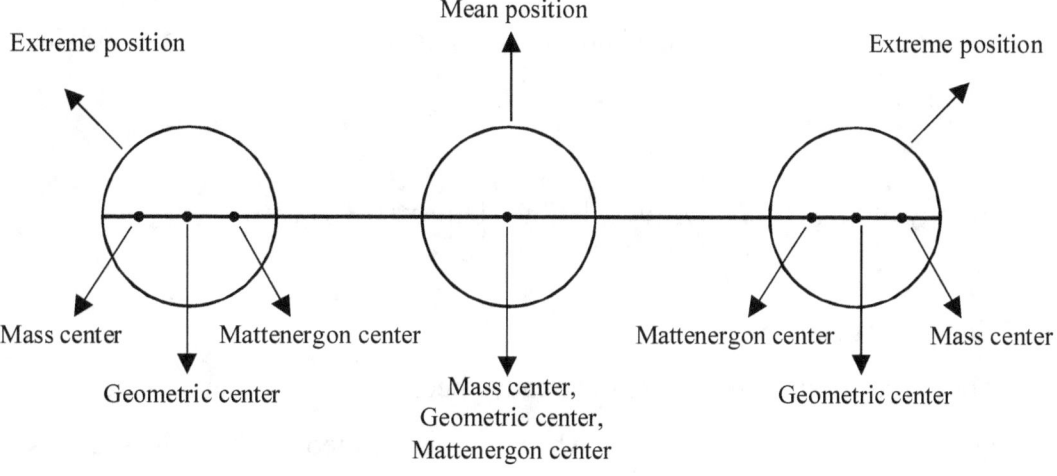

Fig. 54

When the Akashon is at extreme position and starts moving towards the mean position, the mass center is shifted in a direction opposite to the direction of restoring force when the Akashon is at mean position no force acts, therefore mass center, geometric center and mattenergon center overlap. While deducing energy of S.H.O the mattenergon quantum mechanic restriction should be kept in view. When the Akashon starting from extreme position reaches the mean position after completing one fourth of external oscillation, the mattenergon center while executing internal oscillations inside the sphere of tranquillity should overlap with the mass center at the external mean position, it is possible if during the time Akashon completes one fourth of external oscillation, the mattenergon center completes $1/4^{th}$, $3/4^{th}$, $5/4^{th}$, ……….. of internal oscillations. If v_m is the frequency of oscillation of mattenergon center and v is the frequency of external oscillations, the following relation satisfies the above condition:

$v_m = (2n + 1)v$

The above relation also satisfies the condition that mattenergon center position should remain unchanged when it again returns to the original position.

Energy of S.H.O $= E = \frac{1}{2}hv_m = \frac{1}{2}(2n + 1)hv = (n + \frac{1}{2})(hv)$

Zero point energy:
The zero point energy has already been discussed under the topic quantization of kinetic energy where it was pointed out how it is useless

energy. It has been shown that zero point energy comes into existence when akashon is set into motion and is dissipated when akashon is brought to rest hence it is also named as resistance energy because it is associated with the resistence that operates during change of state it is very easy to easy to understand it. One process creates it while the other process robs the horded energy. During SHO motion the akashon repeatedly experiences states of rest to motion and motion to rest hence the zero point energy is repeatedly accumulated and consumed during the above mentioned state changes. It is created during the inertial time needed to set into motion the akashon from state of rest to motion and dissipated when akashon changes from state of motion to rest. Zero point energy participates only in the internal process of akashon. One process creates it while the other process consumes it. The zero point energy does not participate in the external process or surrounding. Any attempt to tap it for other external processes than for the internal processes for what it is meant is fruitless. Zero point energy has been much misunderstood actually; zero point energy as seen above is the energy due to internal oscillations of mass center and mattenergon center inside the sphere of tranquillity. Even if the external oscillations stop, internal oscillations are there that is why energy left is said to be zero point energy. Internal oscillations are different from the external oscillations in the sense that while during external oscillations the geometric center changes position, during internal oscillations mass center and mattenergon center oscillate with geometric center as the mean position with frequency equal to the external frequency, hence

Zero point energy = minimum energy = $\tfrac{1}{2}h\nu$

Suppose an oscillator starts loosing energy in stages as required by quantum mechanic restrictions, with each stage of loss of energy, amplitude of the oscillator will decrease, gradually the state will be reached when amplitude of the oscillator will become equal to or less than 2r. r is the radius of the sphere of tranquillity of the Akashon acting as oscillator. So long as amplitude is greater than 2r, the geometric centre of the Akashon must change position during oscillations or Akashon must move in the conventional sense but when amplitude becomes less than 2r there is no need for the geometric centre to shift position. At this stage the mass centre oscillates around the stationary geometric centre as mean position. The above mentioned mode of oscillations with minimum amplitude less than 2r is associated with energy called zero point energy, its value is ½($h\nu_0$) as already seen.If the force applied to the S.H.O. is equal to or less than the inertial time interval, the inertial mattenergon center starts oscillating along with mass center inside the sphere of tranquillity with amplitude less than 2r without the Akashon changing position or having motion in the conventional sense.

The zero point energy can not be tapped because it is not transferred to region of perturbation.When the Akashon is at rest in the conventional sense, the oscillations are there but the energy is confined to the sphere of tranquillity and hence tapping out of this energy is impossible. The oscillations are there in the special or lower state without the Akashon as a whole changing position because force has acted for period less than inertial time interval.The zero point energy is trapped inside the sphere of tranquillity during special type of oscillations. It is bad news for the free lunch advocates becausetapping out of zero point energy is not so simple. In general when the Akashon starts moving away from the mean position in direction opposite to the direction of the restoring force, theinertial mattenergon starts deteriorating and each time a newinertial mattenergon is created when the Akashon after reaching the extreme position just starts moving towards the mean position. The zero point energy is used in the formation of new inertial mattenergon as the Akashon reaches the extreme positions twice in each oscillation. The zero point energy cannot be used for other purpose that is why zero point energy is called useless energy,to tap it is only a wild fantasy. The excess energy apart form the zero point energy is accumulated in the next state of vibrations when the mass and geometric centers changes positions during S.H.O. In the lowest state vibrational motion is confined to sphere of tranquillity, therefore, energy is stored as inertial mattenergon and hence energy is useless because it is only the energy stored in the region of perturbation which can be exchanged, the energy in the region of perturbation is accumulated when the Akashon moves as a whole along with the geometric center. The energy in the region of perturbation is stored as mattenergy in the temporary region of perturbation.

The energy of the oscillator has been derived without using any complex mathematical calculations which appeal to the mind because to arrive at such a simple result no circuitous path is expected and also the concept of zero point energy is clear without any bizarre explanations.**The oscillating Akashon can be located beyond the extreme position contrary to the classical concept.** The above observation is not difficult to understand.

Evolution of Physical Laws

When the Akashon is at the extreme position, it is momentarily at rest and is attracted towards the mean position with maximum restoring force which shifts the mass center away from the geometric center or beyond the

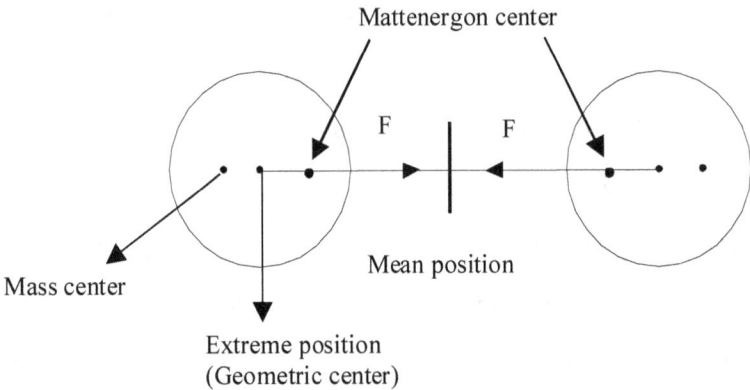

Fig. 56

classical extreme position coinciding with the geometric center after time interval equal to the inertial time. The above observation shows that Akashon can be located beyond the classical amplitude. It can happen so because Akashon is not merely a point as in the classical concept.

THE AKASHON (ETHERON) MOVING IN THE CIRCULAR PATH AND RIGID ROTATOR:

If the Akashon behaves like a point mass and is moving in a circular path, it will be associated with only one type of energy, i.e. rotational energy but actually Akashon has got sphere of tranquillity or sphere of uncertainty having finite dimensions where mass center can be located. When the

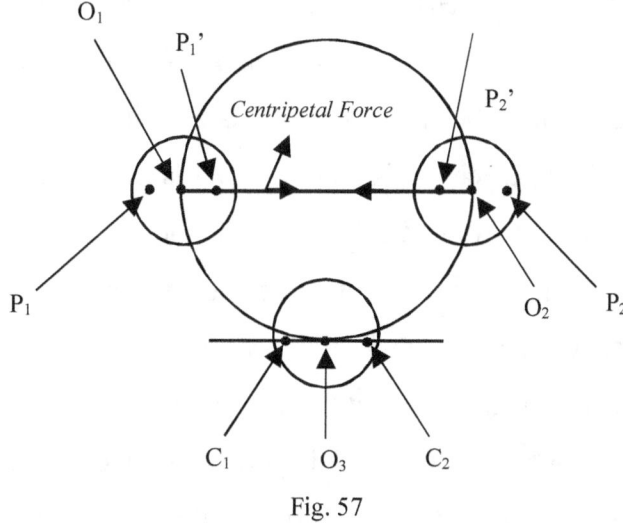

Fig. 57

Akashon is moving in a circular path, the centripetal force directed towards the center of the circle acts on the Akashon. The mass center of the Akashon is directed in a direction opposite to the centripetal force.

The geometric center of the Akashon lies on the circumference of the circle as shown in the diagram at three points O_1, O_2 and O_3. Suppose Akashon is at point O_3 and centripetal force is withdrawn, the Akashon will move in a direction tangent to the circle at O_3 with uniform velocity. The above process will not be instantaneous because readjustment requires time equal to inertial time interval. The energy associated with the Akashon due to tangential velocity can be called as tangential rotational energy. O_3 is the position of geometric center and C_2 is the position of tangential inertial mattenergon which oscillatealong the direction of tangent to the circular path at point O_3 when Akashon moves tangentially to the circular path after the withdrawl of centripetal force. O_1 and O_2 are the geometric centers lying on the circumference, P_1 and P_2 are the locations of the mass center. These two points are situated diamctrically opposite to each other. When the Akashon is at point O_1, the mass center P_1 is shifted towards left of the geometric center and when the same Akashon is at point O_2 after traveling semi-circular path, the mass center is shifted towards right of the geometric center. From above observations it is easy to conclude that as the Akashon complete half revolution, the mass center executes half oscillation or when the Akashon completes one revolution, the mass center executes one simple harmonic oscillation having geometric center as the mean position, therefore rotating Akashon is associated with internal oscillations of the mass center inside the sphere of tranquillity. Finally it is concluded that as the Akashon moves in a circle, it is associated with tangential motion and internal simple harmonic motion oscillation because oscillations take place inside the sphere of tranquillity. The two types of motion mean two types of inertial mattenergon and hence two types of energies.Thetwo types of motions hence two types of energies are not possible if Aakashon behaves like point particle and has got no extended dimensions contarary to the real picture. In that case only rotational energy is possible.

(1) Rotational Energy (tangential velocity):

If ω is the angular velocity, the Akashon after completing one circle in time $2\pi/\omega$ return to the original state, therefore, the centre of tangential velocity mattenergon should remain unchanged, and this is possible if during time $2\pi/\omega$, themotionmattenergon centre executes complete number of oscillations.

The rotational frequency = $\omega/2\pi$

The frequency of rotational inertial mattenergon = $\nu = l\omega/2\pi$, $l = 0, 1, 2, 3, ..$

If v is the tangential velocity which is also equal to the velocity of mattenergy waves

$\omega = v/R$, $l = 0,1,2,3, ...$

$\nu = lv/2\pi R$, $\lambda = h/mv$, $v = h/m\lambda$, put value of v

$\nu = lh/2\pi Rm\lambda$

As already seen, mattenergy wavelength and velocity are related as

$\nu\lambda = v$, $\nu = v/\lambda$

Again putting value of ν,

$v/\lambda = lh/2\pi Rm\lambda$

$mvR = lh/2\pi$

The above equation shows that angular momentum mvR is quantized. Bohr assumed the above relation as such while calculating the energy of hydrogen atom. Here the above relation has been proved and it will be used further.

Evolution of Physical Laws

$\lambda = h/mv$ or $mv = h/\lambda$

$2\pi r = l\lambda$

$\nu = \omega/2\pi = lv/2\pi R$,

$mvR = lh/2\pi$

$v = lh/2\pi mR$, putting the value of v

$\nu = l^2h/4\pi^2 mR^2$

Rotational energy = ½hν = $l^2h^2/8\pi^2 mR^2$

(2) Simple Harmonic Oscillator Energy

As already discussed in addition to rotational energy Simple Harmonic Oscillator energy is also associated with the rotating Akashon due to internal oscillations. The time required to complete one revolution is $2\pi/\omega$ and during this time, center of the inertial mattenergon associated with S.H.O. should also complete one oscillation. The frequency of S.H.O must always be equal to rotational frequency $\omega/2\pi$ because the radius of rotator is fixed, frequency is also fixed.

The rotational frequency = oscillating mattenergon frequency

As calculated above $\omega/2\pi = lh/4\pi^2 mR^2$

Mattenergon oscillaiting frequency $\nu' = lh/4\pi^2 mR^2$

Energy due to internal oscillation of S.H.O. = E_{SHO} = ½hν' = $lh^2/8\pi^2 mR^2$

Total energy is got by adding up the two contributions of energy

Total energy = $(l^2h^2/8\pi^2 mR^2) + (lh^2/8\pi^2 mR^2) = l.(l+1).h^2/8\pi^2 mR^2$

Moment of inertia = $I = mR^2$

Total energy = $l.(l+1).h^2/8\pi^2 I$

It is seen that the energy of rigid rotator is sum of two energy terms, the first energy is due to rotational mattenergon and second energy term is due to internal S.H.O motion associated with the mass center of the rotating Akashon.

The above relation gives the rotational energy of single Akashon moving in circular path of radius R. While discussing the relation for the energy of Akashon having circular motion by using the simple method (not Schrodinger wave equation). **It is seen for the first time that the Akashon is subjected simultaneously to two types of motion circular and simple harmonic motion. That is why the total energy is the sum of energies associated with circular motion and simple harmonic motion. The two types of motions are possible because the mass packet is not point like in nature as assumed in conventional treatments. The Akashon has got extended or finite size.** There is no physical explanation that why the energy of rotor is the sum of two energy terms if Schrodinger wave equation is used to get the energy value According to mattenergon quantum mechanics it is so because total energy is the sum of rotational energy and SHO energy.

The simple harmonic oscillations are dependent on the rotational motion therefore, as soon as rotational motion stops, S.H.O. also disappear or it can be said that whole energy of S.H.O. is useful.

Let us next find out the rotational energy of two Akashons (mass packets) separated with distance R and the distance of Akashons from the axis of rotation are r_1 and r_2. The masses of Akashons are m_1 and m_2. Suppose ω is the angular velocity

$R = r_1 + r_2$

$m_1 r_1 \omega^2 = m_2 r_2 \omega^2$

$m_1 r_1 = m_2 r_2,$ $\qquad\qquad r_2 = R - r_1$

$m_1 r_1 = m_2(R - r_1)$

$r_1 = m_2 R/(m_1+m_2)$ $\qquad\qquad r_2 = m_1 R/(m_1+m_2)$

Total moment of inertia $= I = \quad m_1 r_1^2 + m_2 r_2^2$

$I = (m_1 m_2 / m_1 + m_2).R^2$

Total energy of the rotator $= 1/2(m_1 r_1 \omega^2) + 1/2(m_2 r_1 \omega^2)$
$\qquad\qquad = 1/2(\omega^2 m_1 r_1 R)$
$\qquad\qquad = 1/2(\mu R^2 \omega^2)$

Reduced mass $= \mu = m_1 m_2 /(m_1+m_2)$

The above expression shows that the energy of rigid rotator consisting of two mass packets is equal to the energy of single particle having mass equal to the reduced mass and revolving in circular path of radius R which is equal to the distance of separation of the two Akashons or the mass packets. Thus it is seen that the rigid rotator consisting of two Akashons can be replaced with single Akashon.

Angular momentum:

Energy of rigid rotator = $l(l+1)h^2/8\pi^2 I$

$I = \mu R^2$

Energy of rotator = $E = L^2/2I$

Here L is the angular momentum

$L = \sqrt{l(l+1)}.h/2\pi$

It is seen that the angular momentum is quantized. The angular momentum is vector quantity and its direction follows the thumb rule. The direction of angular momentum is perpendicular to the plane of rotation of the mass packet.

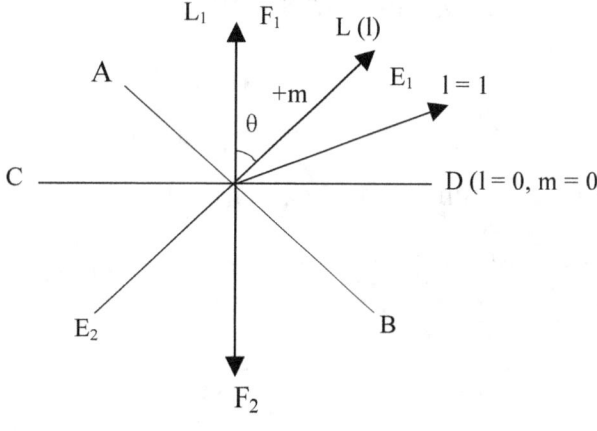

Fig. 58

Suppose AB the plane perpendicular to the paper is the plane of rotation, therefore, total angular momentum is directed along OE_1 which is perpendicular to the plane AB, the component of the total angular momentum along OF_1 is L_1 and CD is the plane of rotation of the hypothetical Akashon because the actual Akashon is rotating only in the plane AB.

$L_1 = L\cos\theta$

The particle rotating in the plane CD is only hypothetical particle and no real force acts on it, therefore, no force mattenergon is associated with it. The energy of the hypothetical Akashon rotating in the plane CD is due to its tangential motion. Tangential motion only without any S.H.O hence, it has got rotational kinetic energy only.

Rotational kinetic energy, $E_1 = m^2 h^2 / 8\pi^2 I'$

$E_1 = L_1^2 / 2I'$

$L_1 = mh/2\pi$,   m like l is an integer

Total angular momentum $L = h/\{2\pi \sqrt{l(l+1)}\}$

L_1 is component of total angular momentum along an arbitrary fixed direction.

$\cos\theta = L_1/L = m/\sqrt{l(l+1)} = (m/l)/\sqrt{\{1+(1/l^2)\}}$

The total angular momentum can be directed along OE_1 or OE_2 depending on the direction of motion of the Akashon

$\cos\theta \leq 1, m/\sqrt{\{l(l+1)\}} \leq 1$, and $m \leq \sqrt{(l^2+1)}$

Put l = 0 and m = 0, the possible value of m which satisfies the above relation is 0. When l = 1, m = 0, +1. When l = 2, the possible positive values of m applicable to the above relation are 0, +1, +2, similarly when l = 3, the positive values of m are 0, +1, +2, +3. m is associated with the angle θ which is the angle between the total angular momentum and component of the angular momentum along fixed direction say Z axis. The direction of the total angular momentum is reversed with the reversal of the direction of rotation, therefore the sign of the component angular momentum along Z-axis is also reversed which means that for each positive value of m, there is negative value.

When, l =1, m = -1, 0, +1;
       l =2, m = – 1, -2, 0, +1, +2;
       l = 3, m = -3, -2, -1, 0, +1, +2, +3.

In general for the given value of l, (2l + 1) values of m are possible, therefore, (2l + 1) directions of the total angular momentum are allowed. The orientations of angular momentum are also quantized. l and m are the angular and magnetic quantum numbers associated with hydrogen atom. These two quantum numbers have been derived very easily without using SchrödingerWave equation involving complicated solution of differential equations. l gives the total angular momentum while m is associated with the component of total angular momentum in arbitrary direction.

The energy relation for the Akashon moving under different conditions have been derived without using abstract mathematics which is required when Schrödinger wave equation is used to solve the energy problems. All the energy problems can be solved very easily without using the concept of wave function. Even high school student without any knowledge of long differential equations can solve the problem.

Spin Rotational Energy of Akashon:

Spinning Akashon is equivalent to rotating Akashon. If it is supposed the whole mass of spinning Akashon is associated with the equivalent Akashon rotating in radius r, r is the radius of sphere of tranquility of spinning Akashon. Two types of frequencies or mattenergons are associated with spinning Akashon ν_m is the mattenergon frequency due to rotation and ν is the S.H.O frequency of the shell center located in the surface of sphere of tranquillity. ν is given by the following relation,

$4\pi r\nu = c$

r is the radius of sphere of tranquillity of Akashon

So that mattenergon quantum restriction may be obeyed two frequencies are related as under:

$\nu_m = 2n\nu$ or $\nu_m = (2n + 1)\nu$

n = 0, 1, 2, 3, ...

Spin energy = $E = \frac{1}{2}h\nu_m$

If, $\nu_m = (2n + 1)\nu$

$E_1 = \frac{1}{2}.h\nu(2n + 1)$

If $v_m = 2nv$, $E_2 = 2hv.(n/2)$

$v = c/4\pi r$

$c = h/4\pi mr$

m is here the mass of the Akashon

Putting the value of v and c in energy relation, and $mr^2 = I$

$E_1 = (h^2/8\pi^2 I)(n/2 + 1/4)$

$E_2 = (h^2/8\pi^2 I).(n/2)$

$E_1 = (h^2/8\pi^2 I)(n/2 + 1/4)$ ………………….. (1)

To make the above energy relation like rotational energy relation

$E = (h^2/8\pi^2 I).s.(s+1)$, put $(n/2 + 1/4) = s(s + 1)$

$(2n + 1) = 4s.(s + 1)$

s is the spin quantum number

From 1st relation, when n = 1, s = 1/2; n = 7, s = 3/2; n = 17, s = 5/2; n = 31, s = 7/2

For calculating suitable value of s, the values of n is given by,

$s = \{-1 + \sqrt{(2n +2)}\}/2$

If second relation is used $n/2 = s(s + 1)$, when n = 4, 12, 24, 40 the values of s are 1, 2, 3, 4, ….. respectively. The suitable value of n is associated with s, $s = \{-1 + \sqrt{(2n +1)}\}/2$. Only that value of n are allowed which give whole number value to the square root factor.

The same relation as given above can be used to find out the next value of n. the fraction value of spin are applicable to Fermion, while whole number values of are applicable to Bosons.

$E = (h^2/8\pi^2 I)(s)(s+1)$

Spin angular momentum = $L_s = \sqrt{(E^2/2I)} = (h/2\pi)\sqrt{\{s(s+1)\}}$

s is spin quantum number. Its value as already seen is 1/2, 3/2, 5/2, ….. or 1, 2, 3, 4, 5, ……

The spin quantum number has been deduced by the application of mattenergon quantum mechanics.

$v_m = 2nv$, when $n = 1$

$v_m = 2v$

The frequency of spin rotation is twice the oscillation frequency of the shell, which leads to the conclusion that the spinning Akashon will attain the same position after two rotations.

Unification of general relativity and quantum mechanics:

The general relativity is based on the equation $r = r_0\sqrt{1 - v^2/c^2}$. The same equation is applicable while calculating the value r for microscopic particle and for black hole when r → 0. The paradigm that different theoretical tools are needed to study the behaviour of microscopic particles (quantum mechanics) and of massive objects (general theory of relativity) is negated it is so because two theories are bases on the same above mentioned equation.

The same equation leads to the derivation of de-Broglie wavelength equation which is used for deriving basic energy relation of mattenergon quantum mechanics. By using the basic energy realtion the energy problems of elementary particles can be exactly solved without using Schrodinger equation and thus disproving THE GENERAL PARADIGM THAT CLASSICAL MECHANICS CAN NOT BE USED TO SOLVE ENERGY PROBLEMS OF ELEMENTARY PARTICLES. It is seen that the two theories are unified because the foundation of the two theories is the same fundamental equation.

If the radius 0f sphere tranquillity is large the akashon is microscopic and the same akashon behaves like black hole when radius of sphere of tranquillity $r \rightarrow 0$ the laws of nature are not expected to change with mere change in the radius of sphere of tranquillity. Contrary to the paradigm that classical mechanics cannot be used to calculate energy of microscopic particles mattenergon quantum mechanics is successful to get accurate energy results by using classical mechanics.

The mass packet akashon has got extended dimensions. Theoritically the akashon presence extends up to infinity and that is why during entanglemend law of locality is not violated and spooky action at distance is possible and this reslolves the main objection.Regarding Einstein's view that God does not play dice the mattenergon quantum mechanics is not based on probability, moreover the model of mass packet the akashon is such that it can be considered to be having both point and extended size. The mass centre of akashon is located inside the sphere of tranquility. The radius of sphere of sphere of tranquillity of electron is of the order of 10^{-13} m which means that error in locating the mass centre can not be more than above magnitude which is nearly zero. The radius of sphere of tranquillity is very small.for heavy mass to which general theory of relativity is applicable hence the error in locating the mass centre is even much more less and hence akashon can be located at a point Even when the akashon is moving the region of location does not extends from $+\infty$ to $-\infty$ as is supposed in mainstream quantum mechanics even then the location point lies inside the nearly point like sphere of tranquillity.

HYDROGEN LIKE ATOM:

The problem is to find put the energy of hydrogen atom. The solution of Bohr was very simple. The Schrodinger wave equation requires very difficult higher mathematics. The value of energy got is negative. The proposed simple thought experiment shows that the energy of hydrogen can not be negative. Suppose stationary electron is at distance r from the proton. The energy required to move the electron to infinity will be $(1/4\pi).(e^2/r)$. The energy as calculated from Schrodinger wave equation is $(1/2\pi).(e^2/r)$. The less value of actual energy shows that electron is not in stationary state in the hydrogen atom. The electron is in state of motion, therefore, kinetic energy is released when it is brought to rest. The extra released energy leads to less value of ionization energy. The above experiment shows that kinetic energy is always associated with the hydrogen atom. Kinetic energy is always positive, it can not be negative. The mattenergon quantum mechanics used to calculate the energy dispenses with the negative energy concept.

Hydrogen atom is discussed by using mattenergon quantum mechanics. The problem is solved very easily without using Schrodinger wave equation. Hydrogen atom is formed by the combination of electron and proton initially separate by infinite distance. Suppose at any instant electron is at distance r from the nucleus carrying charge equal to Ze.

Fall in electrostatic potential energy = $- (1/4\pi\epsilon_0).(Ze^2/r) = - KZe^2/r$

$K = 1/4\pi\epsilon_0$

Next it will be discussed into what forms of energies the electrostatic potential energy liberated as above is converted. As the electron is accelerated towards the nucleus starting from rest electromagneton is formed and with the formation of electromagneton instantaneous miniscule residual charge is developed. The additional miniscule charge gives knee jerk additional acceleration to charge which leads to the expulsion of photon in direction making an acute angle with the direction of motion of electron. The expulsion of photon changing the straight path of electron to circular orbit around the nucleus.

The energy problem of hydrogen atom can be very easily calculated by using the results of rotational energy of Akashon. There is no need to use Schrodinger wave equation and very complex mathematical solution. The important result of rotational Akashon as derived by mattenergon quantum mechanics used here is $mvR = lh/2\pi$, R is the radius of orbit of rotation. The relation was proved and not taken as such as in case of Rutherforde model. In case of rigid rotator R is fixed, but in case of hydrogen atom electron is rotating around the positive nucleus and radius of orbit can change. In case of rigid rotator minimum value of $l = 0$ but this is not allowed in case of hydrogen atom because if $l = 0$, R is 0 which is not possible because if R approaches 0, electrostatic potential energy approaches infinity hence minimum value of $l = 1$.

Replaceing l by n where n = 1, 2, 3, 4 ....
The electrostatic force of attraction = KZe^2/R^2
Centrifugal force = mv^2/R

Equating two forces, $mv^2/R = KZe^2/R^2$

$R = KZe^2/mv^2$

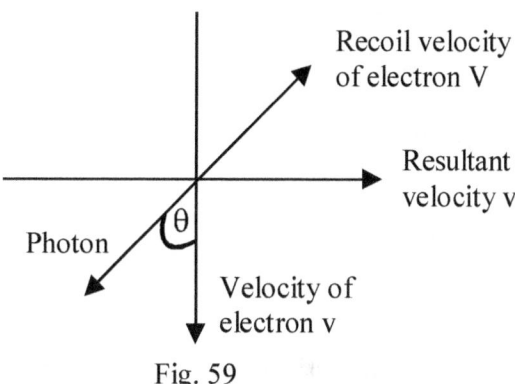

Fig. 59

Suppose recoil velocity imparted to electron is V due to emission of photon. The electron which is moving towards proton with velocity v to form hydrogen atom has got two velocities. If the resultant of two velocities is v in direction perpendicular to the initial direction of motion of electron, the electron will start orbiting around the proton to form hydrogen atom

From the diagram it can be seen $v = \sqrt{v^2 + V^2 + 2V\cos(180 - \theta)}$

Squaring both sides, $V = 2v\cos\theta$

$$\cos\theta = \frac{V}{2v}$$

This equation gives the direction of emission of photon, only this direction of emission of photon and recoil velocity will put the electron in desired trajectory to form hydrogen atom.

The following result has been proved while discussing energy of rotator and has not been assumed as such as was done by Bohr.

$mvR = nh/2\pi$

$v = nh/2\pi mR$

$v^2 = n^2h^2/4\pi^2m^2R^2$

Putting the value of v,

$R_n = n^2h^2/4\pi^2me^2KZ$

It shows that the radius of rotation is quantized. During the formation of hydrogen atom electron starts from infinity towards proton. In the initial state energy is zero. Energy liberated during formation process when electron is at distance R_n from proton E_n-o = fall in electrostatic potential energy = $- KZe^2/R_n$

E_N is the energy of the sysrem after the formation of hydrogen atom

Putting the value of R_n as deduced above $E_n = (- 4\pi^2me^4K^2Z^2)/n^2h^2$

The above value of energy based on energetic or as calculated by using mattenergon quantum mechanics as calculated below by using mattenergon quantum mechanics is twice the value as calculated by using Schrodinger equation. It shows that half of the liberated potential energy is retained by the hydrogen atom. It is obvious that half of the liberated kinetic energy will be taken up by the electron as kinetic energy which proves without any doubt that contrary to the quantum mechanics paradigm the electron is not stationary in hydrogen atom it moves in well defined orbits in hydrogen atom. BOHR WAS NOT WRONG. The main objection to his theory was that the electron will fall into the nucleus due to loss of energy because it is subjected to acceleration directed in direction perpendicular to direction of motion hence there is no effect of acceleration on velocity The relation for electromagneton energy which is precursor of electromagnetic radiation shows unless there is change in speed there is no emission of radiation hence electron is not expected to collide with the nucleus. While deriving energy relation Bohr assumed quantization of angular momentum without any proof but by applying Mattenergon quantum mechanics to the energy solving problem of rigid rotor it was proved easily without using any complicated mathematics that angular momentum is quantized.

The above radius value can be used to solve the radiation energy problem of hydrogen atom, as the electron reaches from infinity to distance R from proton potential energy released is equal to Ke^2/R, there is no need to put negative sign before it because it is simply released energy as in case of wound up spring energy. Where does the released energy go!

Kinetic Energy = ½(mv^2), put the value of mv^2 as calculated earlier,
Kinetic Energy = ½(Ke^2/R)
The above relation shows that the half of the potential energy released os changed to kinetic energy, while the other half is radiated as electromagnetic radiation. The total electromagnetic energy lost when the electron reaches the nth orbit is equal to ½(Ke^2/R_n),

$R_n = n^2h^2/4\pi^2me^2K$

Total energy lost = $2\pi^2 me^4 K^2/n^2 h^2$

E_r is the value of total radiation energy lost by the time electron reaches the nth orbit. The energy lost is in steps as the electron drops from higher orbit to lower orbits which is associated with the value of n. The above treatment shows that the source of electromagnetic radiation of hydrogen atom is the electrostatic potential energy of the electron, while Schrodinger wave equation solution is silent about the source of energy. The same result can also be got by using concept of temporary electromagneton energy. It has been seen that electromagneton energy is twice the kinetic energy half of which is radiated out as kinetic energy while the other half is retained as kinetic energy.

The following method of energy calculation is based upon electromagneton concept because rotating particle is charged. Charge center of the rotating electron is attracted by the positive nucleus hence it moves away from the geometric center. As discussed during rotationalenergy, mass center completes one oscilation during one rotation similarly charge center completes one oscilation during one revolution of electron. During one complete rotation charge center also completes one oscillation according to restriction imposed by first principle of mattenergon quantum mechanics, charge center must be at the same position when it reaches the starting position after completing any number of rotations.

Unlike the rigid rotator R incase of hydrogen atom is variable, hence charge center acting like mass center can have multiple values of frequency but frequency must be whole number multiple of rotational frequency.

If ω is the angular velocity, rotational frequency = $\omega/2\pi$

The oscillation frequency of charge center = $\omega/2\pi$

As calculated during rotation of Akashon $\nu = n\omega/2\pi = nh/4\pi^2 mR^2$,

Putting value of R, $\nu = 4\pi^2 me^4 K^2 Z^2/n^2 h^2$

The charge center here is undergoing internal oscillations,

therefore, $E_m = \frac{1}{2}h\nu = 2\pi^2 me^4 K^2 Z^2/n^2 h^2$

E_m is not the energy of electron or hydrogen atom. It is the total electromagnetic radiation lost by the electron when it reaches the nth state.

The oscillation of charge center creates electromagneton, the electromagneton at once changes to photon. The total electromagnetic radiation emitted when the electron is at n_2^{th} stage after starting from infinity,

$En_2 = 2\pi^2 me^4 K^2 Z^2/n_2^2 h^2$

The total radiation emitted when it is at n_1^{th} stage is $En_1 = 2\pi^2 me^4 K^2 Z^2/n_1^2 h^2$

Suppose $n_2 > n_1$ the electromagnetic radiation emitted when electron steps down from n_2 to n_1 state, $\Delta E = \{2\pi^2 me^4 K^2 Z^2/h^2\} \cdot \{(1/n_1^2) - (1/n_2^2)\}$

The same amount of energy has to be supplied when the electron steps up from n_1 to n_2 state. The hydrogen atom is formed by the combination of proton and electron initially separate by very large distance as the electron just start moving toward nucleus in straight path as already discussed straight path changes to circular orbit. As soon as the electron starts rotating charge center is subject to oscillations and oscillating charge emits photon, the recoil velocity of emitted photon leads to the fall in the orbit radius because $mvR = nh/2\pi$ or electron is pushed inwards. In the next orbit again the above process is repeated and the electron goes on loosing radiation stepwise till the lowest step when $n = 1$ is reached. The total energy radiated out when it reaches the lowest step is $E = 2\pi^2 me^4 K^2 Z^2/h^2$.

Minimum distance from the nucleus $R = h^2/4\pi^2 me^2 KZ$

Combining the second relation with the energy relation it can be easily shown that $E = KZe^2/2R$

The above relation proves that out of the total electrostatic potential energy liberated half of the energy is changed to electromagnetic radiation. Thus, it is seen that source of electromagnetic radiation is electrostatic potential energy. It has also been seen above that $mv^2 = KZe^2/R$.

Kinetic energy $= \frac{1}{2}mv^2 = KZe^2/2R$

The above relation proves that half of the electrostatic potential energy is converted to kinetic energy of the electron. Due to kinetic energy of electron it is seen that electron is very much in rotational motion around the nucleus. The motion imparts dual nature to the electron. When the hydrogen atom energy is calculated by the application of Schrodinger wave equation, E is said to be energy of the hydrogen atom and its sign is negative but here when energy is calculated by using mattenergon quantum mechanics the energy E is the total electromagnetic energy radiated by the electron when it reaches the special point after starting from infinity.

Relation between l and n:

It has been seen that half of the electrostatic potential energy is changed to rotational kinetic energy of the electron while the other half is radiated out as electromagnetic radiation when the electon reaches the special point.

It will now be seen that actually half of the electrostatic potential energy is changed partially into rotational kinetic energy. Suppose for calculating the rotational energy quantum number n is used.

Rotational energy $E_{rot} = n(n+1)h^2/8\pi^2 mR^2$

Half of electrostatic potential energy $= E_{elc} = Ke^2/2R$, where $K = 1/4\pi\epsilon_0$

$E_{rot}/E_{elec} = \{n(n+1)h^2/8\pi^2 mR^2\}\{2R/Ke^2\} = n(n+1)h^2/8\pi^2 mRKe$

Putting the value of R as deduced earlier

$R = n^2 h^2/4\pi^2 me^2 K$

$E_{rot}/E_{elec} = n(n+1)/n^2 = 1 + (1/n) = 1$

Hence above result is wrong, therefore another quantum number will have to be used to calculate the rotational energy, suppose that is l

Evolution of Physical Laws

$E_{rot} = l(l+1) = h^2/8\pi^2 mR^2$

If, $E_{rot} = E_{elec}$

$(l^2 + l) = n^2$

$l = -1 \pm \sqrt{\{(1 + 4n^2)/2\}}$

$n = 1, 2, 3, \ldots$

If above equation is used by putting value of n, l will not be whole number hence,

$E_{rot} \neq E_{elec}$

E_{rot} cannot be $> E_{elec}$

The only possibility is that $E_{rot} < E_{elec}$ or $l(l+1) < n^2$

$n > \sqrt{\{(l^2 + l)\}}$

The above relation shows that the whole of the half of the electrostatic potential is not changed into rotational energy.

From the above relation when $n = 1, l = 0$; $n = 2, l = 1$

In general, for any value of n, $l = 0, 1, 2, 3, \ldots (n-1)$

l is angular quantum number

The relation between l and m has already been deduced. Therefore, all the four quantum numbers n, l, m and s has been deduced by using mattenergon quantum mechanics. Quantum number s was deduced while discussing roatational energy. The above result shows that for the given value of n, there are n values of l or n values of rotational energy are permitted or there are n multiple orbits for the rotation of electron. The orbit with less value of l has less radius of rotation which leads to the conclusion that it is easy to remove the electron from the path with greater value of l. depending on the value of l the paths of rotation can be named as when l = 0, 1, 2, 3, …… the paths are s, p, d. In wave quantum mechanics the paths are named as orbitals.

The angular momentum of electron when l = 0 and residual energy:
As already discussed $E_{rot}/E_{elec} = l(l + 1)/n^2$
Rotational kinetic energy $E_{rot} = \{l(l + 1)/n^2\}E_{elec}$
As discussed earlier $E_{rot} < E_{elec}$

It means that whole of the half of electrostatic potential energy is not changed to rotational energy. The unused electrostatic potential energy can be called as residual energy ($E_{elec} - E_{rot}$). The residual energy forms a sort of energy envelope in the form of electromagneton around the nucleus or proton. As the electron starting from infinity moves toward the proton the energy of the electromagneton envelope goes on increasing. The residual energy is used in the formation of bonds.

Half of the electrostatic potential energy = Residual energy (Electromagneton) + Kinetic energy

More residual energy is there for lesser value of l. When l=0 rotational kinetic energy=0 hence residual energy=half of the electrostatic potential energy=maximum residual energy.

The greater the value of l the greater is the value of kinetic energy of electron. Therefore, though the electrons belong to the same orbit (n) the orbitals s (l = 0), p (l = 1), d (l = 2), …….. have not got same kinetic energy. Kinetic energy of s is 0. The order of kinetic energy is s < p < d < f. The above relation shows that s, p, d, f electrons though having the same value of n are associated with different kinetic energies. According to mattenergon quantum mechanics the electrons are rotating in the orbits. The electron does not exist as electronic cloud as proposed by quantum wave mechanics.

When l = 0 rotational kinetic energy E_{rot} = 0, which leads to the conclusion that electron should stop rotating but it is not allowed. It can be said that in the special case when l = 0 the electromagneton or residual energy is maximum and it changes to hybrid state and behaves like electromagneton energy or kinetic energy hence the electron does not come to stop. Wave quantum mechanics is silent about the source of electromagnetic radiation emitted by hydrogen atom but here when hydrogen atom is discussed by using mattenergon quantum mechanics it is seen that source of radiation is electrostatic potential energy of the electron. Moreover, the source of bond energy the residual energy is also electrostatic potential energy.

Lamb's Shift:

When n = 2, the electron can be in two degenerate states 2s, 2p. But more residual energy is associated with the 2s electron than 2p electron which makes these two orbitals non-degenerate. Energy is liberated during transition of 2s to 2p. This is called lamb's shift. According to QED energy difference is due to vacuum interaction involving virtual particles which does not seem so plausible. The residual energy is not only the cause of the lamb's shift but it also leads to the formation of co-valent bond.

Formation of co-valent bond:

When two hydrogen atoms are brought nearer to each other repulsive force is expected because outer part of the atoms carries electronic charges according to wave quantum mechanics, therefore, like charges should repel each other but contrary to this attractive force there which leads to the formation of co-valent bond. As discussed the nuclei of each atom are surrounded by envelope of residual energy or electromagneton. The elcetromagnetons belonging to different atoms are attractive in nature, hence, there is attractive force between the atoms. As atom come more nearer to each other the same charge carried by the nuclei start exerting repulsive force, at a certain distance the attractive force of electromagnetons and repulsive force become equal. At this stage co-valent bond is formed.

Co-valent bond of H_2^+:

In this molecule only one atom is associated with residual energy packet which is shared between the two atoms hence bond is formed.

Uncertainity in locating the electron:

According to the Schrodinger wave equation, Ψ^2 is the probability of locating the electron in hydrogen atom. Mattenergon energy is associated with the electron and its presence is needed for the location of electron. Partially empty electron does not loose its identity. It is not spread out in the form of electronic cloud. It is the mattenergy envelope which is spread out. Electron is not so fragile that it is easily pulverized to form infinite number of shells of which it is composed of and the shells are scattered around the nucleus and when the electron islocated the shells regroup to form compact electron. It is against the second law of thermodynamics. The orbitals merely give the patten of distribution of the mattenergy.

Maximum number of electrons in an orbit:

The energy radiated out by the maximum number of electrons allowed in an orbit should be equal to the maximum electrostatic potential energy. Maximum electrostatic potential energy is equal to $4\pi^2 me^4 K^2 Z^2/h^2$. Suppose maximum number of electron allowed in orbit is equal to the n, the energy liberate is equal to $2\pi^2 Nme^4 K^2 Z^2/n^2 h^2 = 4\pi^2 me^2 K^2 Z^2/h^2$.

Hence, $N = 2n^2$

Hydrogen atom and new form of Schrodinger Wave Equation:

Schrodinger wave equation is not rigorously applicable to hydrogen atom due to following reasons:

1. The Schrodinger wave equation is applicable to the motion of single particle. If the mass center of the electron and proton is considered to be equivalent to single particle, the origin of the coordinate system will be shifted away form the nucleus proton because the origin will have to be mass center around which proton and electron are rotating.

2. Hydrogen atom is formed by the combination of electron and proton initially separated by infinite distance.

   Electron + Proton → Hydrogen atom + Electrostatic Potential Energy

   The above equation shows that after the formation of hydrogen atom, electron energy should be negative.

3. According to mattenergon quantum mechanics as applied to hydrogen atom. While solving Schrodinger equation the value of electrostatic potential is calculated accurately by pin pointing the position of electron but later on while discussing wave function it is stressed that wave function gives only the probable position and hence electron exact position cannot be known. How to explain the paradox.

Electrostatic potential energy (V) = Radiated electromagnetic energy (E_m) + Kinetic energy of electron. There is no need to put negative sign before V because it is just like the energy of wound up spring.

Kinetic energy = $V - E_m = -(E_m - V)$

Schrodinger wave equation can be derived very easily by using relation, Kinetic Energy = $p^2/2$, $\lambda = h/p$ and then differentiating the wave disturbance equation

$$(\partial^2 \Psi/\partial x^2) + (\partial^2 \Psi/\partial y^2) + (\partial^2 \Psi/\partial z^2) + (8\pi^2 m/h^2)(\text{Kinetic energy})\Psi = 0$$

According to wave quantum mechanics Schrodinger wave equation is written as

Evolution of Physical Laws

$$(\partial^2\Psi/\partial x^2) + (\partial^2\Psi/\partial y^2) + (\partial^2\Psi/\partial z^2) + (8\pi^2 m/h^2)(E - V) = 0$$

If the above equation is solved the sign of E is found to be negative and E is taken as the energy of hydrogen atom. The energy E as calculated from the above result leads to negative energy value hencehydrogen atom is expected to be stable but atom is highly unstable and is stabilized by combining with other hydrogen atom. Schrodinger wave equation leads to the wrong result and need to be modified.According to mattenergon quantum mechanics E_m is the energy radiated out by the electron when it reaches the specified point. It is not the energy of hydrogen atom.

$$(\partial^2\Psi/\partial x^2) + (\partial^2\Psi/\partial y^2) + (\partial^2\Psi/\partial z^2) + (8\pi^2 m/h^2)\,(\text{Kinetic Energy}) = 0$$

Put value of kinetic energy as deduced by the application of mattenergon quantum mechanics, the equation is modified as given below,

$$(\partial^2\Psi/\partial x^2) + (\partial^2\Psi/\partial y^2) + (\partial^2\Psi/\partial z^2) - (8\pi^2 m/h^2)(E_m - V) = 0$$

When the above equation is solved while keeping the sign of V as positive because according to mattenergon quantum mechanics V is just energy like the energy of woundup spring,the value of E_m is found to be same but with positive sign and the rest of solution is not changed.For solving above equation no imaginary terms has to be used.

THE RELATION BETWEEN THE ORBITAL QUANTUM NUMBER 'l' AND PRINCIPLE 'n':

The relation between n and l can also be calculated very easily. As already seen the energy is emitted by the electron as it moves towards the nucleus when it is located at distance R = 2nr. The recoil velocity of the emission of photon puts the electron in an orbit given by the following equation. If R' is the radius of the orbit and λ is the wave length,

$l\lambda = 2\pi R'$

R' is not equal R. R' is slightly less than R because with the emission of photon, the required velocity of photon pushes the electron inwards making R' < R. If the photon is emitted at n^{th} point, the electron follows the orbital path with radius between n^{th} and $(n + 1)^{th}$ point

$R = 2nr = nh/2\pi mc$
$2\pi R' = l(h/mv)$
$l/n = (R'/R)(v/c)$
$R'/R < 1$
but v / c is much less than one, therefore, l < n
n = 1,2,3,……………..   l = 0,1,2,3,………….

The interpretation of the result when l = 0:

In the ground state of hydrogen like atom when n = 1, the value of l = 0. when l = 0, it leads to the conclusion that there is no angular moment or there is no orbital motion of the electron which further leads to the strange conclusion that if the electron is not moving where does it lie in state of rest. The electron is at rest when l = 0, therefore, the Schrödinger wave equation should not be applicable because Schrödinger wave equation is relevant when the mass packet is in motion.

One explanation for the zero angular momentum in the ground state while retaining the motion of the electron is proposed in the conventional treatment by projecting the orbital motion of the electron in symmetric orbits around the nucleus, but it is contrary to the established laws because if the electron in motion is to change its path, force is required and there is no source of special force according to the conventional treatment. The value of magnetic quantum number 'm' has already been found while discussing rigid rotator.

The state of hydrogen like atom can be easily visualized when discussed in the light of the special treatment which does not use Schrödinger wave equation for the calculation of energy etc. According to the present treatment when l = 0, the sphere of tranquillity of the electron envelopes the nucleus or in other words the nucleus lies snug inside the sphere of tranquillity of the electron which is possible because the radius of sphere of tranquillity of the electron is much bigger than that of nucleus, therefore, there is no orbital motion of the electron but spins motion is possible.

When the electron initially present at infinite distance from the nucleus starts moving towards the positively charged nucleus, the common sense predicts that ultimately the electron should collide with the nucleus in the same manner as an object thrown from height comes to rest on reaching the surface of the earth, but actually it does not happen so, because when the electron is at certain points from the nucleus, the photon carrying extra self-electrical potential energy, shoots out of the electron which results in changing the straight path of the electron into circular orbit in the same manner as satellite is put into desired orbits by the booster rockets. The points at which photons are emitted is given by $R = 2nr$. If all the electrons moving towards the nucleus as the common sense predicts drop in the nucleus, there would have been no creation of the elements with characteristic properties endowed by the electron moving around the nucleus in various orbits. These tiny electrons are tireless workers and are ever busy to make the universe colorful.

SPIN ANGULAR MOMENTUM OF THE AKASHON OR THE MASS PACKET:

The spin angular momentum can be determined by applying the principle of inverse variation. That the principle of inverse variation is applicable to spin angular momentum is discussed in detail later on.

$4\pi r L_s$ = constant

L_s and r_0 the spin angular momentum and radius of sphere of tranquility. The constant on RHS has got dimensions of length x spin angular moment.

Planck's constant h has got spin angular momentum dimensions, therefore constant = $r_0 h$

$4\pi r_0 L_s = r_0 h$

$L_s = (½)(h/4\pi)$

If length on R.H.S. is put equal to n.r,

$4\pi r.L_s = n.r.h$

n = 1, 2, 3, ……..

$L_s = nh/4\pi$

Putting values of n,

$L_1 = ½ (h/2\pi)$, $L_2 = h/2\pi$, $L_3 = 3/2 (h/2\pi)$, $L_4 = 2(h/2\pi)$

The odd and even values of n give spins of fermions and bosons respectively.

The other explicit method of calculating spin angular moment is given below.

The angular momentum of the spherical shell is required to calculate the angular momentum. The angular momentum of spherical shell of radius r having mass m and rotating with angular velocity ω is given by the following equation

$L_0 = (2/3) m r^2 \omega$

The Akashon or the mass packet can be visualized as composed of infinite number of concentric spherical shells rotating with equatorial linear velocity which follows the principle of inverse variation. The Akasha is gliding past the fixed displaced shells in the same manner as a stream is flowing over the pebbles. As already expressed, the cumulative effect of the circulating Akasha and the fixed oscillating shells of the Akasha is illusion of the mass. If v is the equatorial linear velocity and r is the radius, the application of the principle of inverse variation gives the following result

$4\pi r v = k_v = $ constant

$v = r\omega \qquad 4\pi r^2 \omega = k_v$

$\omega = d\theta / dt \qquad 4\pi r^2 (d\theta/dt) = k_v$

$(1/2) r^2 (d\theta/dt) = dA$

dA is the area swept by the radius of the equatorial circle in dt time interval. Putting the value of dθ

$(8\pi) dA / dt = k_v$

$dA / dt = k_v / 8\pi = $ constant

Evolution of Physical Laws

the above result is the Kepler's second law associated with planetary motion. The above law is derived by using the principle of inverse variation.

$$L_0 = (2/3)mr^2\omega = (2/3)mrv$$

Suppose dm is the mass of the concentric spherical shell having radius equal r

$$L_0 = dL = (2/3)dm.rv$$
$$m = h/(4\pi rc)$$
$$dm = -\{h/(4\pi r^2 c)\}.dr \qquad v = k_v/4\pi r$$
$$dL = -k_v.(2/3)(h/16\pi^2)(dr/r^2)$$

$$L_S = \int_0^{L_S} dL = -(2/3)(k_v h/16\pi^2 c). \int_\infty^{r0} dr/r^2 = (2/3)(k_v h/16\pi^2 r_0 c)$$

L_S is the total spin angular momentum

From the above equation it is seen that $4\pi r_0 L_S$ = constant

Thus, it is seen that the principle of inverse variation is applicable to the total angular momentum also.

Angular Momentum from Principle of Inverse Variation:

$4\pi r.L_S$ = constant = Length x Angular Momentum

Planck's constant h has dimensions of angular momentum and is also universal constant. The constant on R.H.S. can be put equal to r_0h.

$4\pi r.L_S = r_0 h$

r_0 is the radius of sphere of tranquillity

$L_S = h/4\pi = ½ (h/2\pi)$

The angular momentum can also be derived in another also,

$4\pi r_0 v = k_v$

Put the value of k_v

$L_S = (h/6\pi).v/c$

v is the equatorial linear velocity of the innermost spherical shell forming the surface of the sphere of tranquility. The rotating mass shell present in the surface of sphere of tranquillity has got two frequencies. Frequency v_0 is intrinisic frequency due to shell being part of the Akashon. The value of intrinisic frequency v_0 is got by applying the principle of inverse variation. The second frequency is due to rotation and its value is equal to the rotational frequency as already discussed while examining the rotational energy. The rotational frequency is v. It is thus seen that center of mass shell is subjected to two frequencies simultaneously in the same manner as S.H.O. experiencing two oscillations, oscillation of the mass center when the Akashon as a whole is subjected to S.H.O. and oscillation of mattenergon center. In the present case natural frequency of the shell can be equated to basic frequency of S.H.O. The basic or natural frequency is

v_0 and its value depends on position of the shell. The frequency of shell due to rotation can be equated to mattenergon frequency v, which is variable. If the mattenergon quantum mechanics condition is applied the oscillating shells will be in the same position if the rotational and intrinisic shell frequencies are related as given below:

$v = (2n + 1).v_0$ or $v = 2nv_0$

v is odd or even multiple of v_0

The marked shell of the Akashon spinning like a top will be in the same state when it returns to the initial state after completion of whole number of rotations if above condition is followed.

$v_0 = c/4\pi r_0$

$v = v/2\pi r_0$

v is tangential velocity

Put these values in the frequency relation,

$v/c = (2n+1)/2$ or $v/c = 2n/2 = n$

As already deduced the spin angular momentum $= (h/6\pi)(v/c)$

If spinning electron is considered as one particle. Actually there are three spinning particles. The intrinisic mass of the electron forms one particle, the second contribution comes from the mass equivalents of two energy packets associated with the electron as electromagneton. The third contribution is of the mattenergon though the mass equivalent mattenergon is very very less but it has been seen that spin angular momentum is independent of mass.

The total angular momentum $L_S = (h/6\pi)(v/c).3 = (h/2\pi)(v/c)$

The above result can also be derived by considering spinning Akashon as tried of three Akashons each having $h/6\pi$ as spin angular momentum.

Put value of v/c

$L_S = (h/2\pi)(2n + 1)/2$

The above expression gives spin of fermions

n = 0, 1, 2, 3, …………………..

$L_{S0} = (1/2)(h/2\pi); L_{S1} = (3/2)(h/2\pi);$…………………………..

The spin quantum numbers are $\pm 1/2, \pm 3/2,$ …………………..

When v/c = 2n, $L_S = nh/2\pi$

n = 1, 2, 3, 4, ……..

$L_S = h/2\pi, 2h/2\pi, 3h/2\pi,$ ……..

The expression gives spin of Bosons

The equatorial velocities of the surface of the sphere of tranquillity are got from $v = \{(2n + 1)/2\}.c$

The lowest velocity is when n = 0

$v_0 = c/2; v_1 = 3c/2$ ……………….

The value of velocity more than c does not violate any rule because while discussing the terminal velocity, it was seen that it is the mass center which cannot move with velocity greater than velocity of light. In case of spinning Akashon the mass center is stationary while only surface is moving.

Evolution of Physical Laws

Let us discuss special case $v = 2nv_0$, v here is the rotational frequency when $n = 1$, $v = 2v_0$. If $v = 1$, $v_0 = 1/2$ which means that after the completion of one rotation the mass shell completes only half of the oscillation. Therefore, the mass shell or Akashon as a whole does not return to the initial state after the completion of one rotation or motion through 360^0 as is expected. The Akashon will return to the initial state only after the completion of two rotations.

The correct spin quantum number has been derived by using pure classical mechanics and considering the Akashon like spinning top which invalidate the paradigm that to quantum mechanical particles classical mechanics is not applicable. In the following diagram the electron is represented like spinning top.

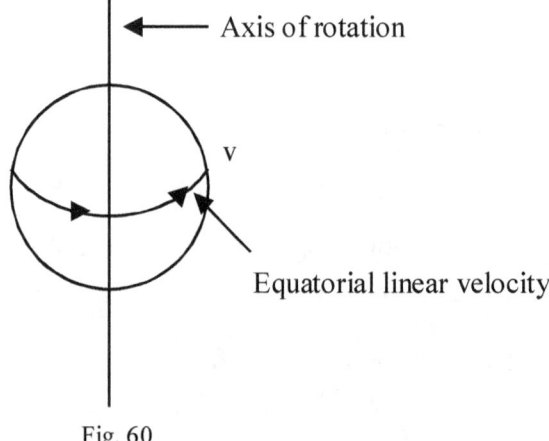

Fig. 60

The magnetic moment of spherical shell having charge and angular velocity equal to dQ and ω respectively is $(2/3).dq.\omega.r^2$. The charge packet is composed of infinite number of rotating concentric spherical shells.

$d\mu_S = (2/3)dQ.r^2.\omega = (2/3)dQ.rv$

$4\pi rQ = k_Q = $ constant

$dQ = -(k_Q / 4\pi r^2).dr$

$4\pi rv = k_v$

$v = k_v / 4\pi r$

$d\mu_S = -(2/3)(k_Q / 16\pi r^2).dr / r^2$

$$\mu_S = \int_0^{\mu_S} d\mu_S = -(2/3)k_Q / 16\pi^2 \int_\infty^{r_0} dr/r^2 = (2/3)k_Q k_v / 16\pi^2 r$$

$4\pi r_0 \mu_S = $ constant

SPIN MAGNETIC MOMENT OF THE CHARGED MASS PACKET:

The magnetic moment can also be determined by applying the principle of in verse variation.

$4\pi r \mu_s = $ constant

The above relation has been just proved in the previous derivation. The constant has got dimensions of length x magnetic moment, therefore, it is apt to put constant equal to r.(Qh/m). Q and m are the charge and mass of the Akashon. The factor (Qh/m) has got dimensions of magnetic moment.

$4\pi r \mu_s = r(Qh/m)$

$\mu_s = (Qh/4\pi m)$

The other explicit method of calculating μ_s is given below.

The above result shows that like the angular momentum, magnetic moment can also be calculated by the application of the principle of inverse variation. Put values of $k_Q = 4\pi rQ$, $k_v = 4\pi rv$ and $r_0 = h/4\pi mc$ in above relationship for μ_s.

$\mu_s = (Qh / 6\pi m).(v/c)$

as in case of angular momentum put $v/c = (2n+1)/2$

$\mu_s = (Qh/6\pi m).\{(2n+1)/2\}$

Put n = 1

$\mu_1 = (Qh/4\pi m)$

The spin magnetic moment of electron is got by putting n' = 1 as in case of angular momentum. The spin magnetic moment relation is derived for those mass packets in which the radius of the sphere of tranquillity of the mass packet is taken equal to the radius of the sphere of tranquillity of the charge packet. In case of magnetic moment of proton, the radius of the sphere of tranquillity of the mass packet is much less than that of charge packet, therefore, the spin magnetic moment of the electron and proton, is expected to be not same though the magnitude of charge is same.

Anomalous Magnetic Moment of Electron:

The practical magnetic moment of electron is a little more than $Qh/4\pi m$. Instead of charge value Qcharge can also be represented as e. For the extra value of spin magnetic moment of electron there should be some extra charge. Let us see what is the source of the charge. The centripetal force due rotation of electron introduces extra strain which results in shifting away of the charge shells. The shifting away of the charge shells leads to the development of extra charge. The mass is not affected because mass is dependent on the oscillation of the mass shells and not on the strain or shift. The extra charge in strain charge if r_m is the mass radius of sphere of tranquillity of electron the value of the half of the circumference, πr_m of the circle of radius r_m is measure of strain because it is associated curvature. Strain is inversly proportional to length of curve because charge shells present in straight line has to be strained to change it into curve form. Strain charge is proportional to strain. Hence extra stain charge $e^\wedge \alpha 1/\pi r_m$. The charge on the eletron $e\alpha 1/r_0$, r_0 is the charge radius of the electron.

From above relations it is concluded that $e^\wedge/e = \pi \frac{}{r_0 \; r_m} = \alpha/\pi$ α is fine structure constant. Additional charge is accommodated on mass m, therefore, extra magnetic moment $= e^\wedge h/(\pi)4\pi m$. Putting the value of extra charge$^\wedge$ and converting the additional magnetic moment into B.M. units it can be easily seen g factor increases by α/π unit. Actually the additional charge is developed as doublet of charge packet with each part of the doublet rotating in the opposite direction, each part of doublet being of opposite charge the contribution to the magnetic moment is added up but contribution to the total charge is zero because doublet is neutral.

THE FUSION OF TWO AKASHONS:

Suppose two Akashons with radii of spheres of tranquillity r_1 and r_2 fuse together to from an Akashon with radius of sphere of tranquillity equal to r. Let us discuss what is the relation between the masses of the Akashons before and after fusion by assuming that either the total surface area of the spheres of tranquillity before fusion is equal to the surface area of the sphere of tranquillity formed by the fusion or the same relations is there between the volumes of the spheres of tranquillty before fusion and after fusion.

(1) Suppose surface area remains same –

$4\pi r_1^2 + 4\pi r_2^2 = 4\pi r^2$

$r_1^2 + r_2^2 = r^2$

$(r_1 + r_2)^2 = r_1^2 + r_2^2 + 2r_1r_2$

$r_1 + r_2 = \sqrt{(r^2 + 2r_1.r_2)}$

divide by r_1r_2

$(1/r_1) + (1/r_2) = (1/r_1r_2).\sqrt{(r^2 + 2r_1r_2)}$

$(1/r_1) + (1/r_2) = (r/r_1r_2)\sqrt{\{1 + (2r_1r_2/r^2)\}}$

If m_1 and m_2 are the masses of the Akashons undergoing fusion and m is the mass of the Akashon formed after fusion, therefore, the following relation is applicable

Evolution of Physical Laws

$$(1/r_1) + (1/r_2) \propto m_1 + m_2$$

$$1/r \propto m$$

$$\{(1/r_1) + (1/r_2)\} / (1/r) = (m_1 + m_2)/m$$

$(m_1 + m_2)/m$ is called mass ratio

$$(m_1 + m_2)/m = (r^2/r_1r_2)[1 + \{(2r_1r_2)/r^2\}]^{1/2}$$
$$= \{(r_1^2 + r_2^2)/r_1r_2\}\{1 + (2r_1r_2/r^2)\}^{1/2}$$
$$= [\{(r_1 - r_2)^2 + 2r_1r_2\}/r_1r_2][2 - \{(r_1 - r_2)^2/r^2\}]^{1/2}$$
$$= [2 + \{(r_1 - r_2)^2/r_1r_2\}][2 - \{(r_1 - r_2)^2/r^2\}]^{1/2}$$

The mass ratio increases as the difference $(r_1 - r_2)$ of the radii of the Akashons participating in the fusion increases, it is so because with large difference in the value of radii, the first factor increases much more as compared to the fall in the magnitude of the second factor and this brings about over all increase in the mass ratio. The minimum value of mass ratio is got by taking Akashons of equal radii $r_1 = r_2$

$$(m_1 + m_2)/m = 2.\sqrt{2}$$

For Akashons with unequal radii

$$(m_1 + m_2)/m > 2.\sqrt{2}$$

The above result shows that the fusion of Akashon always leads to loss of mass or evolution of energy. It is also seen that the Akashons with unequal radii lose more mass on fusion

Evolution of Physical Laws

(2) When the volume of the sphere of tranquillity is taken to be constant –

$(4/3)\pi r_1^3 + (4/3)\pi r_2^3 = (4/3)\pi r^3$

$r_1^3 + r_2^3 = r^3$

$(r_1 + r_2)^3 = r_1^3 + r_2^3 + 3r_1r_2(r_1 + r_2)$

$(r_1 + r_2)^3 = r^3 + 3r_1r_2(r_1 + r_2)$

$(r_1 + r_2) = r[1 + \{3r_1r_2(r_1 + r_2)\}/r^3]^{1/3}$

$\{(1/r_1) + (1/r_2)\} / (1/r) = (r^2 / r_1r_2)[1 + \{3r_1r_2(r_1 + r_2)\}/r^3]^{1/3}$

$(m_1 + m_2)/m = [(r^2 / r_1r_2)][1 + \{3r_1r_2(r_1 + r_2)\}/r^3]^{1/3}$

Let us next see the value of factor $r^2 / r_1 r_2$

$(r_1 + r_2)^3 = r_1^3 + r_2^3 + 3r_1r_2(r_1 + r_2) = r^3 + 3r_1r_2(r_1 + r_2)$

divide by r_1r_2

$(r_1 + r_2)^3 / r_1r_2 = (r^3 / r_1r_2) + 3(r_1 + r_2)$

$r^2 / r_1r_2 = [(r_1 + r_2)/r].[\{(r_1 + r_2)^2 / r_1r_2\} - 3]$

it can also be seen that

$(r_1 + r_2)/r = [1 + \{3r_1r_2(r_1 + r_2)/r^3\}]^{1/3}$

$r^2 / r_1r_2 = [1 + \{3r_1r_2(r_1 + r_2)/r^3\}]^{1/3}.[1 + \{(r_1 - r_2)^2 / r_1r_2\}]$

From above, it is seen that $(r^2 / r_1r_2) > 1$

If $r_1 = r_2$, $(m_1 + m_2) > (7)^{1/3}$

The above result leads to the conclusion that $\{(m_1 + m_2)/m\} > 1$. Therefore, there is loss in mass or evolution of energy on fusion of Akashons.

UNIO (ONION) – MODEL OF NUCLEUS:

This model unites all the models of nucleus. The nucleus of an atom is supposed to be composed of neutrons and protons held together by strong nuclear force. The observed properties of the nucleus are explained by independent particle model or liquid drop model of the nucleus. No single model of the nucleus can explain satisfactorily all the properties of the nucleus. The protons and neutrons are not present as such in the nucleus because if it is so the nucleus is expected to produce protons and neutrons profusely on being struck by strong projectile. The new model of the nucleus presented here can explain the expected behaviour of the nucleus. The properties of the nucleus associated with the independent particle model and drop-like model can be explained with the single model presented here.

According to the model, the atomic nucleus is composed of number of Akashons having common mass center or in other words, the spheres of tranquillity of the Akashons are concentric. The innermost Akashon is the Centron. The concentric Akashons around the Centron are termed as the Concentrons. The mass associated with each Concentron is much lessas compared with the mass associatedwith the Centron. The total number of Concentrons enveloping the Centron plus one is equal to number of protons and neutrons in the conventional sense or mass number. In the diagram, the boundaries of the spheres of tranquillity of the Centron and the Concentron are depicted.

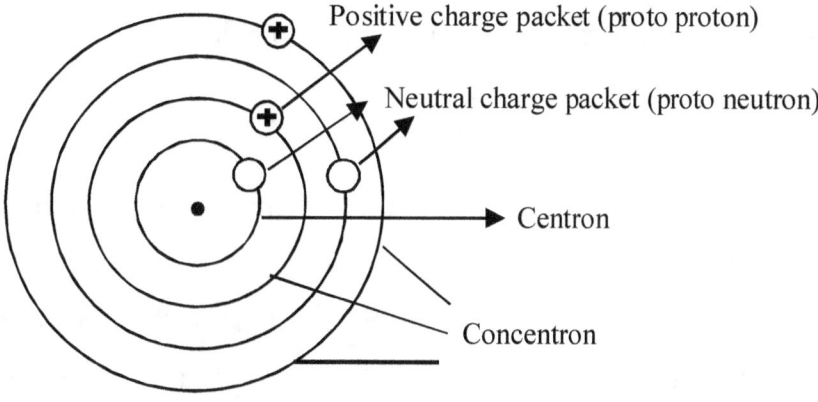

Fig. 61

By volume as given below means the volume of sphere of tranquillity. If the volume got by subtracting volumes of spheres of tranquillity of two consecutive concentrons is same as the value of volume of the sphere of tranquillity of the Centron, the following relation is possible. If the value of the radius of the outermost concentrons is R and A is mass number, r is the radius of sphere of tranquillity of the centron, by applying the above conclusion, the following result is obtained.

$(4/3)\pi R^3 = (4/3).\pi r^3.(A)$, r is the radius of the sphere of tranquillity of the centron; $R = r.A^{1/3}$

The Centron is an Akashon having mass less than proton. If a proton is true Akashon, the radius of the sphere of tranquillity of proton = $h/4\pi nc$ = 1.05×10^{-16}m. The Centron has got mass less than proton therefore r is greater than 1.05×10^{-16}m, the radius of sphere of tranquillity of proton. Similarly the conventional radius of nth Concentron or radius of nucleus is greater than $1.05 \times 10^{-16} \times A^{1/3}$ m. The experimental value of radius of nucleus is equal to $1.4 \times 10^{-15} \times A^{1/3}$ m.

Due to comparatively large value of the sphere of tranquillity of the outermost Akashon, the mass density of the sphere of tranquillity is relatively less, therefore, R can be put equal to the radius of the nucleus. The Centron can share mass with the Concentron. The constant k or r_0 involved in the radius relation is not strictly constant, because as more and more Concentrons are being added the mass of the Centron increases hence, the radius of the sphere of tranquillity of the Centron with which k is associated will decrease, so its magnitude goes on slightly decreasing with increase in the value of A which results in shorter value of the radius of the particular Concentron when it is belonging to nucleus with high value of A as compared to the Concentron associated with the nucleus with low value of A.

The nucleus contains two types of charge packets, the positive charge packet and the neutral charge packet. The neutral charge packet is formed by the combination of positive and negative charge packets. The positive charge packet is termed as proto-proton and the neutral charge packet as proto-neutron because of the conversion into protons and electrons respectively under suitable nuclear reaction. Proto-protons and proto-neutrons differ from the conventional protons and neutrons in the sense that the masses are not equal to conventional protons and neutrons. Proto-proton and proto-neutrons are inter-convertible in the nucleus. Usually the outermost Concentron prefers to accommodate the proto proton. Because of the special positions of the proto proton and proto neutron, unbalanced cosmic force directed towards the common center is acting on proto proton and proto neutron and under the influence of the centripetal force the proto proton and the proto neutron are rotating around the Centron. The proto protons and proto neutrons have also spin motion. The spin and the orbital angular moments of nucleus are the resultants of moments associated with the proto proton and proto neutron. Similarly, magnetic moment of the nucleus is the resultant of magnetic moments of proto protons.

The less dense region of perturbation of the proto proton and proto neutron are composed of pro energy or mattenergy shells that is why the proto packets can glide easily through the Akashon shells while having rotational or spin motion. The mass equivalent of proto packets is less than the free proton or free neutron because the region of perturbation of proto packets is composed of mattenergy shells. The above observation leads to the conclusion that the mass of the nucleus is less than the sum of the masses of proto packets and free neutrons and protons, which are formed, from proto packets present in the nucleus under suitable conditions. The cosmic force acting on mattenergy shells is directed towards the center of proto packets. The centers of the proto packets are located on the surface of the spheres of tranquillity of the concentrons, therefore nearly half of the region of perturbation of the proto packets which lies inside the sphere of tranquillity of the given concentrons is less populated with mattenergy shells than the other half which s situated outside as a result of unequal distribution of shells, the net cosmic force is directed toward the centron and the value of force is very high, the resultant strong force supplies centripetal force required for the rotational motion of the proto packets. For the conversion of proto neutron into proto proton, charge doublet of proto neutrons separates and negative part mixes with the proto proton to change into proto neutron and leaving behind proto proton. In the conversion of proto proton and proto neutron indicates that there are no Concentrons with which only proto protons or proto neutrons are to be associated. In atoms with low mass number, only one proto proton or proto neutron is associated but with elements having high mass number, the distribution of proto proton or proto neutron is jumbled as will be seen later on. The mass associated with proto proton or proto neutron is equal to the mass of the Concentron with which proto particle is associated because these proto particles are moving. The unbalanced cosmic force acing on each proto particle is directed towards the centerto supply centripetal force which keeps the proto particles moving in circular path. The source of unbalance cosmic force is presence of more inwards pushing shells on that side which is away from the center than on the side which is near to the center.

The mattenergy shells associated with the proto packets are formed by the conversion of Akashon shells of the Akashon with which the proto packets are associated. The rim of the spheres of tranquillity acts as shell for the accommodation of proto packets. The proto packets which are at more distance from the centron are formed by sucking in large number of Akashon shells of the Akashon with which the proto packet is associated as compared to proto packets which lie nearer to the centron, therefore the masses of the concentrons become less and less as the distance form the centron increases. In other words, the outermost cocentrons are more porous as compared to the inner most centrons. As the distance of the proto packets increases from the centron, the unbalanced cosmic force also become less because with increase in the number of shells associated with the proto packet the disproportional distribution of shells also becomes less, therefore outer proto packets are loosely bound in the nucleus.

In the model of the nucleus as presented here the concept of conventional binding energy is redundant because it is unbalanced cosmic force, which acts as strong binding force. The range of strong cosmic force is very very less which is clear from the following relation:

The Akashon shell density = k_1/r^4
The mattenergy shell density = k_2/r^4

The value of k_2 is very very less as compared to k_1. Therefore the range of the force is very very less. According to the conventional model of the nucleus, it is difficult to imagine how in tiny volume of the nucleus protons and neutrons packed like sardines can rotate and spin at the same time without any collision.

As discussed earlier, with increase in the number of concentrons or mass number the radi of sphreres of tranquillity of outermost concentrons becomes so close together that a band is formed and at certain stage there is fusion of concentrons or Akashons.

As discussed earlier on fusion of Akashon always leads to the liberation of energy and the energy so liberated is used to convert proto packets to full-fledged protons, neutrons and smaller Akashons, which combine together to form particles like Alpha particles. This explains the spontaneous disintegration of the nucleoli with high mass.

RADIOACTIVITY:

As already discussed the radius of sphere of tranquillity of nth, $(n+1)^{th}$ and $(n+2)^{th}$ are given below:

$r_1 = k.n^{1/3}$, $r_2 = k.(n+1)^{1/3}$, $r_3 = k.(n+2)^{1/3}$

The difference in the radii of $(n+1)^{th}$ and n^{th} concentrons

$\Delta r = k\{(n+1)^{1/3} - n^{1/3}\} = k.n^{1/3}\{(1+1/n) - 1\}^{1/3} = k.n^{1/3}\{(1/3n) + ...\}$

The above result shows that as the value of n increases, the difference in the radii of two consecutive concentrons becomes less and less. Therefore, in case of nucleus having high mass number, the surfaces of the outermost concentrons nearly overlap.

It is seen that in case of a nucleus having high mass number, the surfaces of the outermost Concentrons nearly overlap. Moreover, as already seen the radii of the Concentrons associated with nucleus having high value of A or n are relatively squeezed inwards. The two factors result in the nearly merging together of the surface of the spheres of tranquillity of the outermost Concentron to form blurred band. The proto protons and proto neutrons are distributed randomly in the band. The outermost Concentrons are under strain due to overlapping and random distribution of proto protons and proto neutrons. The strain in the outermost Concentrons is the cause of un-stability or radioactivity of the nucleus with high value of mass number. The heavy nucleus becomes stable by the ejection of some particles. The important factor for the disintegration of heavy nucleus is not only charge factor but above observation is very important. In case of heavy nucleus the number of neutron is much more than that of therefore, contribution of charge factor is not of much importance towards unstability of heavy nucleus.

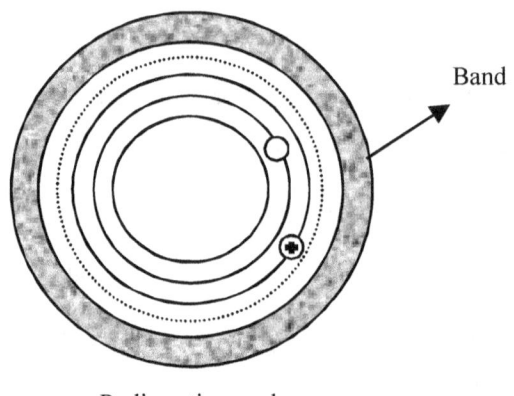

Radio active nucleus

Fig. 62

The blurred band contains proto-proton and proto-neutron in random state in the same manner as molecules are present in the surface of liquid. The fragments which can overcome the central attractive force which keeps the proto particles in centron escape the blurred band to form disintegration products and the rate law is that of first order reaction because only one fragment acquires excess activation energy. Usually heavy radio-active particles are source of β particles but emission of proton is not there. It is so because α particle experience more repulsive force (+2) than proton (+1).

In the blurred band compound particles are created which disintegrate in the same manner as explained while discussing unstable elementary particles here again, coulomb force or weak force helps in disintegration. The particles created in the blurred band and the rest of atoms have eccentric mass centers and therefore separate or disintegrate in the same manner as other unstable compound elementary particles, the distance of separation of mass centers is measure of life of radio-active atom. The greater the distance of separation of mass centers of the particles created in the blurred band and the rest of atom, the shorter is the lifetime of the compound atom.

The nuclei present in the bulk form do not disintegrate simultaneously because each nucleus is in different state of randomness and hence has different life period. The above behavior of the nuclei makes the disintegration process as the first order reaction. There is no need to introduce the concept of binding energy to understand radioactivity when the protons and neutrons are not present as such in the nucleus, therefore, no binding is required.

The region of perturbation of the Centron is limitless, therefore, it embrace all the Concentrons. The Centron is like the spider sitting at the center of the web and as any disturbance in the strands of the web (Concentrons) is instantaneously communicated to the spider, similarly, the Centron is affected by the state of un-stability or disturbance in the Concentron. The Centron is reservoir of mass of the nucleus. The Centron exchanges mass with the Concentrons during radioactivity or nuclear reactions through proto protons and proto neutrons.

The radius of sphere of tranquillity of the nth Concentron $= kn^{1/3}$
The mass of nth Concentron $= h / 4\pi ckn^{1/3} = k' / n^{1/3}$

The above relation shows that the mass of the Concentron decreases with increase in the value of n. The mass of the outer Concentrons of the heavy nuclei is very small, therefore, heavy nuclei are relatively porous to the attack of projectile.

FISSION OF THE NUCLEUS:

As already seen the outermost Concentrons of the heavy nuclei are in unstable state due to overlapping and random distribution of proto proton and proto neutrons. The outer Concentrons are also porous. The density of the nucleus is not same throughout, inner region is more dense. The Concentrons of the heavy nucleus also lie closer to the Centron than that of the light nucleus. The outer Concentrons are nearly overlapping and are not fused but with the help of say thermal neutrons, the process of fusion can be initiated. The fusion of Concentrons can be called as catastrophic happening because when it starts all the Concentrons around the Centron fuse together to form one lump which further undergo fusion with Centron to form big lump which contains proto protons and proto neutrons randomly distributed in small volume. The high energy neutrons cannot start the catastrophic process because of deep penetration to the inner part of the nucleus. The fusion or catastrophic process is very quick and energy is released because as already discussed the fusion of the Akashons always lead to the emission of energy. Due to evolution of energy and the congestion of proto neutrons and proto protons, the relatively big lump of mass is strained or un-stable and, therefore, it usually breaks up into two packets with mini bang and during this breaking up some higher particles and energy are also liberated. The number of proto protons and proto neutrons near the core of the big lump is less, therefore, the core of the nucleus form the small fragment while the rest of the big lump is changed into big fragment. It is seen that the main source of energy in the fission is the fusion of the Concentrons. During the fusion process there is catastrophic breakdown of the nucleus, and after the breakdown the heavy nucleus is reformed as two new nuclei.

As already stated proto neutrons and proto protons are randomly distributed in nearly overlapped outermost Concentrons of the heavy nucleus. If in the outermost overlapping Concentrons, there is preponderance of proto neutrons, the overlapping of the outermost Concentrons to form lump can be easily initiated such as the thermal neutrons. The example is the fission of $_{92}U^{235}$. The collapsing of the Concentrons by fusion is not easy if there is majority of charged protons in the outermost Concentrons as in case of $_{92}U^{238}$. If the proto protons can be pushed inwards the fission of $_{92}U^{238}$ can also become easy. The fission will be easy if the nucleus is placed in a strong magnetic field.

TRANSMUTATION OF ELEMENTS:

When the stable element is bombarded with projectile, the outermost part of Akashon forms a separate particle giving birth to compound particle with the rest of the atom. The compound particle having eccentric mass centers disintegrates in the same manner as the radio-active or unstable compound particles to give rise to new elements.

THE UNIFICATION OF FORCES:

There are so many forces in nature, but the source of all the forces is the same, the cosmic force. There are four forces in nature, (1) gravitational force; (2) electromagnetic force; (3) weak force; (4) strong nuclear force. The unification of these forces can be studied very easily without the introduction of unrealistic concepts or yard long complicated mathematical equations. We have seen what is gravitational force and electromagnetic force. The source of these two forces is unbalanced cosmic force. These two forces have been studied in detail along with the equations for the measurement of the magnitude of these forces. The above observation indicates that the forces are already unified because they originate from the universal cosmic force. However, we can discuss the unification of **gravitational** and **electromagnetic** force.

Einstein continued to work on unified field theory of gravitation and electromagnetism but he became isolated in this research. With four fundamental forces now identified, gravity remains the one force whose unification is problematic. Attempt here is made in the present text to unify gravitational and electromagnetic forces in a very easy manner. The expression for the mass of the super heavy or N Akashon has been derived. It will be shown here that super heavy Akashon is the particle which easily brings about unification of electromagnetic and gravitational field.

Evolution of Physical Laws

Unification of Electrostatic and Gravitational Forces:

Let us take two NAkashons. The mass of each Akashon is square root $\sqrt{(hc/4\pi G)}$. The gravitational force F_G between these two when placed at distance r = Gm_0^2/r^2.

Put the value of $m_0 F_G = (hc/4\pi)(1/r^2)$

The magnitude of the force is independent of gravitational constant G and mass. The magnitude depends purely on distance r.

Next, let us take two pure charge packets,

The pure charge on each charge packet is $\sqrt{(hc\epsilon)}$

The electrostatic force between two charge packets = $(1/4\pi\epsilon).(hc\epsilon/r^2)$ = $(hc/4).(1/r^2)$.

The electrostatic force between two pure charge packets is independent of the permittivity, like the gravitational force the magnitude depends purely on distance r. The above observation shows that the nature of the field exerted by the N Akashon is same as that of pure charge packet or it can be said that N Akashon brings about unification of electrostatic and gravimeteric field.

From the above discussion it is seen that it cannot be distinguished whether the force is between N Akashons and pure charge packets because magnitude is same or it can be said that through N Akashon electric and gravitational field are unified or it can be said that gravitational field associated with the N Akashon is equivalent to electrostatic field. The same thing can be said that the electrostatic field associated with the pure charge packet is of the same nature as that of gravitational field of N Akashon. The above discussion proves that pure mass packet ⇔ pure charge packet. Gravitational field is associated with mass packet while electromagnetic field is associated with charge packet.

Gravitational field ⇔ Electromagnetic field

The same results can be got in modified manner also.

The pure mass $m = \sqrt{(hc/4\pi G)}$

Pure charge $q = \sqrt{(hc\in)}$

On dividing, $q/m = \sqrt{(4\pi G\in)}$

The above equation relates G and $\in$

The above relation relates G and $\in$

The gravitational force between two N Akashons,
$F_G = Gm^2/r^2$
$m^2 = q^2(4\pi G\in)$
put value of m^2
$F_G = (1/4\pi\in).(q^2/r^2) = F_{elec}$

Similarly, $E_{Elec} = (1/4\pi\in).(q^2/r^2)$,

Put $q^2 = 4\pi G \in m^2$

$E_{Elec} = G.m^2/r^2 = F_G$

The association of G, μ_0 and $\in_0$:

The pure mass $m_0 = \sqrt{(hc/4\pi G)}$

$m_0^2 = (hc/4\pi G)$

$c = \sqrt{1/\mu_0.\in_0}$

put the value of c, $G = (h/4\pi m_0^2).\sqrt{1/\mu_0.\in_0}$

The above equation unifies gravitational field and electromagnetic fields because the three constant $G, \mu_0,$ and $\in_0$ are related through super heavy N Akashon m_0. G and $\in_0$ are not the conventional constants. G and $\in_0$ are the values on the surface of spheres of tranquility of pure mass and charge packet.

Pair production and annihilation are also simple examples of unification of electrical and gravitational forces.

1. **Pair Production:**

    The conversion of photon into electron and positron takes place through highly unstable transitory particle. As already discussed,

    Photon $\xrightarrow{\textit{Transitory state}}$ Electron + Positron

Electromagnetic field is associated with photon, gravitational field is associated with electron, positron hence these two fields are interconvertible.

2. **Annihilation of particles:**

Positron and electron on interaction produced two photons through intermediate transitory state as in the above case.

$$\text{Electron} + \text{Positron} \xrightarrow{\textit{Transitory state}} \text{Photon} + \text{Photon}$$

The above equation like the previous equation also proves that gravitational and electromagnetic fields are intervonvertible.

The super heavy N Akashon can be used to discuss unification of electromagnetic field and gravitational field in another way also. The expression for the mass of the super heavy N Aakashon has been derived. Let us see how much energy is carried by the mattenergon associated with the super heavy N Akashon. In first calculations for the mattenergon energy change in gravitational constant at each point in the region of perturbation is considered. The following is the result as already derived in previous pages.

Mattenergon energy = $G.m_0^2/2r_0$

(Self gravitational potential energy)

It is also known that $m_0 = \sqrt{(hc/4\pi G)}$

Putting value of m_0, energy of permanent mattenergon associated with pure mass = $(hc/8\pi r_0)$

No force constants like G etc. are there like in the first expression to pin point the source of energy or in other words the energy can be called as pure energy. If the Akashon is not unique, mattenergon energy is $(Gm_0^2/4).(1/r_0)$. It is not pure energy because it involves G. Only in case of unique or super heavy N Akashon, total gravitational potential energy of mattenergon is independent of G.

Electromagneton energy when change in permittivity is considered has been discussed earlier. The value of electromagneton energy = $(hc/8\pi).(1/r_0)$. From above calculations, it is seen that if the radii of sphere of tranquillity of the super heavy N Akashon and charge packet are same, the gravitational potential energy stored in the mattenergon is equal to the electrical potential energy stored in the electromagneton. It is seen that no gravitational constant or permittivity is associated with the energy stored. This case is only present when pure mass of the Akashon is considered. From above observation it can be concluded that while considering the energies stored up in electromagneton, no distinction is there whether it is gravitational potential energy or electrical potential energy. The source of first energy is gravitational field and that of the second is electrostaticfield.**It is as clear as the day that unification of gravitational field and electromagnetic field**has been brought about in very easy manner. The charge carrying unique or super heavy N Akashon is composed of neither pure mass nor pure electromagnetic and gravitational energy.It is in a special state of matter. Itmay be unstable and explodes to form photon from electromagneton,gravitons from mattenergon, andmore photons from mass energy $E = m_0c^2$or it being unique in nature N particle may be highly stable.

Evolution of Physical Laws

$$\text{Unique or super heavy Akashon (highly unstable) Electrogravitomagneton} \xrightarrow{\textit{breaks up}} \text{Photon (from electromagneton)} + \text{Graviton (from mattenergon)} + \text{Photon (from } E = m_0 c^2\text{)}$$

In unique Akashon, there is no distinction between matter and energy.

We are discussing four types of fundamental forces and their unification. The unification of electromagnetic and gravitational field has been brought about from the first principle as using some simple arguments. The source of electrostatic, magnetic and gravitational fields is the same i.e. unbalanced cosmic force. When the source of force is same, the unification of the forces is automatically there only thin veil of ignorance needs to be rented with some simple manipulation, the way it has been done in the above text. The electrostatic magnetic and gravitational fields are unified in elcetrogravitomagneton. The three fields are directed along x, y, z coordinates.

Evolution of Physical Laws

THE WEAK FORCE:

The weak force is a special type of electrostatic repulsive force which occurs inside the spheres of tranquillity of two charged Akashons.

The weak force is manifested in β-decay, radioactivity and breaking up of unstable elementary particles. Let us now discuss why the neutron is a source of electron while outside the nucleus. What is the reason that electron is loosely bound in neutron. As the neutron exists as proto neutron in the nucleus, the electrical doublet present in it is concentric so it is stable in the nucleus. The cause of stability is also in inter-conversion of proto neutron and proto protons. If sufficient energy is supplied to the proto neutron in the nucleus, negatively charged part changes into electron while positively charged part into proton after gaining suitable masses. The charge centers also become eccentric and are separated to different extents.

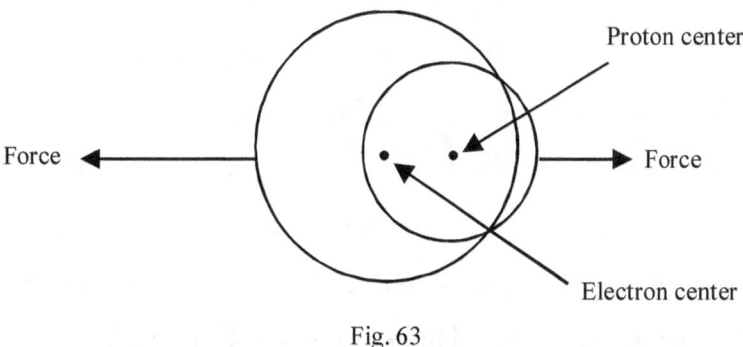

Fig. 63

The charge center of electron lies in the sphere of tranquillity of the proton and vice-versa. Whenever such a situation arises, the two particles which were previously held together are separated. The origin of the force is unbalanced cosmic force which acts outwardly along the shell springs in case of charge packets. In the above example number of outward attracting shells in the left half of the region of perturbation of electron is more than right half because right half is partially occupied by sphere of tranquillity of proton. The disproportionate distribution is the cause of pulling away of the electron towards left from the proton. Similarly, proton is being pulled away towards right due to unbalanced cosmic force. From the above discussion it can be said that source of weak force is the weak electrostatic force or we can say that weak force and electrostatic forces are same. As the electron moves away energy is generated which is stored as neutrino around the electron. The shield of neutrino protects the electron from the attractive force of positive proton therefore, electron can freely leave the proton along with the neutrino. The centers of the neutrino and electron are no longer concentric therefore further on separation of neutrino and electron also takes place. The excess energy generated during separation of proton and electron by electroweak energy is carried by the shells (neutrino) of which the electron is composed of.The emission of neutrino (group of shells) to carry away excess of energy proves that all the elementary particles are composed of shells packed together in special way.How much energy is to be carried by electron, depends on the distance of separation of electron center and proton center. The half life of the neutron and the direction in which electron is to be liberated also depends upon the distance of separation and direction of separation of two charge centers.The above two factors are probability dependent. From the above discussion it is seen that weak force is nothing but weak unbalanced cosmic force associated with the eccentric electrons and protons. The cosmic forces here manifested as electrostatic force.In the next step we are to simply add that weak force and electrostatic forces are same or **are unified**. The weak nuclear force is short range forcebecause it operates only inside the sphere of tranquillity. The weak force is repulsive in nature and hence helps in the separation of fragments. In radioactive disintegration, radioactive atom behaves like compound elementary particles because it breaks into two fragments, one light and other heavy. The light fragment envelopes the heavy fragment in the same manner as electron envelopes the proton during decay of neutron. The process of separation of two fragments is the same as that of neutron. There is no need to include W and Z particles to explain the process.

THE NATURE OF STRONG FORCE:

We have seen that in the unio model of nucleus proton and neutrons are not present as such, however when the proton-neutron, neutron-proton and proton-proton are brought close together, strong force comes into action. It has already been seen that in the charged particles electromagneton and mattenergon envelop the sphere of tranquillity of nucleon. The radii of spheres of tranquillity of mattenergon and electromagneton are equal to the radii of sphere of tranquillity of the charged mass packet and charge packet. The association of electromagneton and mattenergon shells with charge packet shellsand mass packet shells respectively present around the nucleon metamorphosizes the shells to hybrid attractive mattenergon-electromagneton or electromagnetomattenergon. The attractive force is short range force in nature. When the two nucleons are brought very close, there is overlapping of the regions of perturbation of attractive hybrid electromagneton-mattenergon. The charge present on nucleon is repulsive in nature. The range of attractive force exerted by the attractive hybrid electromagneton-mattenergon is less than the repulsive force exerted by charge packets which is long range in nature. The effective range of attractive force exerted by hybrid electromagneton-mattenergon is very very less, of the order of radius of the sphere of tranquillityof proton which is equal to 1.05×10^{-16}m. Actually the attractive and repulsive forces are balanced before the minimum distance is achieved. The nucleon has to be very close together for the operation of the attractive force. As the nucleons are brought nearer and nearer the two opposing forces becomes more and more effective. At a certain distance the opposing forces become equal and the balanced state is reached. On further decreasing the distance, repulsive force due to chargewill operate. The range of force exerted by the attractive hybrid electromattenergon is less than the repulsive force exerted by charge.When the distance increases, the repulsive force becomes less and less in initial stages, therefore, with increase of distance the net attractive force will be there or it can be said that with increase of distance attractive force becomes more.The mode of action of the forces here is the same as electrostatic or gravitational forces with only the difference that the constantsaffecting the force here are large and the range of the force is very very short.

The cause of manifestation of strong force is the same as that of manifestation of gravitational electromagnetic and weak force. The above forces are manifested due to the interaction of regions of perturbation associated with the interacting particles. The strong force is manifested when the regions of perturbation of hybrid electromattenergon interact. The above observation proves that all the forces are unified because the source of manifestation of Special Forces is actually the **Cosmic Force**.

Why the Strong Force is absent when electrons interact. The radius of sphere tranquillity of electron is large as compared to that of neutron and proton. The large radius of sphere of tranquillity of electron also leads to the large radii of spheres of tranquillity of mattenergon and hence mattenergon having less energy is associated with electron. The large radius of sphere of tranquillity of mattenergon is associated with poor density and hence the attractive-interactiveforce of hybrid electromattenergon is very very weak or it can be neglected because it

is not strong enough to overcome the electrostatic repulsive force between two electrons.

Electromagneton energy of electron = 1.89×10^{-16} J

Energy of mattenergon = 2.16×10^{-54} J

The above result shows that mattenergy of electron is much smaller as compared to proton. Therefore, it can be said that attractive force of hybrid electromattenergon is more in case of proton as compared to that of electron or proton neutron are associated with strong attractive nuclear force. The energy of mattenergon associatred with proton is more than that of electron because the radius of sphere of tranquillity of mattenergon associated with proton is less than that of associated with electron.

The strong force here is related to electromagnetomattenergon which are further related to electrostatic and gravitational fields therefore strong force is unified with the electrostatic and gravitational fields. Weak force is already related to electrostatic field. Actually it is the primary cosmic force which is manifested as different forces. Even the chemical combination force between atoms has its cause in the electromagneton which is accumulation of electrostatic potential energy and hence related to cosmic force. It can be easily concluded that the father of all the forces, be it strong or weak etc. manifested in nature is the **Universal Cosmic Force.**

All the above forces have also been unified in an indirect way in the above text. Bringing about unification of force is just showing light to the sun. **No**

strong nuclear force is there between two electrons because due to less mass, the strength of mattenergon and electromagneton energy is lessowing to large radius of sphere of tranquillity of electron as compared to proton and neutron, which results in no strong attractive force. Whatever little attractive force is there, that is overcome by the large repulsive force of similar charges. The radius of sphere of tranquillity of electron and that of proton are 1.93×10^{-13}m and 1.05×10^{-16}m respectively.

The four forces can be explained in nutshell as follows:

1) & 2) Gravitational and Electromagnetic Forces:
The above forces are manifested when the regions of perturbation having infinite range interact. Therefore, range of above forces extend up to infinity

3) Weak Nuclear Interaction:
Weak nuclear forces manifested when special type of electrostatic interaction occurs inside the sphere of tranquillity of two charged Akashons. Unlike the strong force, it is repulsive in nature therefore, it helps in separating the fragments of unstable particles.

4) The Strong Nuclear Force:
The strong nuclear force is manifested when region of perturbation of two nucleons having hybrid electromagneton energy packet and hybrid mattenergon energy packet interact. Like gravitational and electrostatic interaction, this type of force is also expected to be a very long range force.

Evolution of Physical Laws

But on the contrary, it is very very short range force. It is so because it is the interaction of energy packets. Therefore, the effective range lies very close to the sphere of tranquillity. The strong force is also very large in magnitude because it involves the special interaction of energy packets, while in other three forces no such interaction is present.

It has already been discussed that all the above forces are simple manifestation of the cosmic force and hence all the four forces are unified.

THE HUMAN CONSCIOUS:

The human conscious is atomic in nature. The Akashons are gross while the Prakashons are subtle in nature but the packets of human consciousness are still more subtle. The packets of human consciousness can be called as mysterious force packets. The human conscious can also be called as Chetnon from Hindi word "cheta" for human conscious. The Chetnon or conscious force packets are present everywhere but these happen to be more Concentered in the living beings. The Chetnons interact with shells of the Akashon. The negative Chetnon destroy or disturb the orderly arrangement of the shells of the Akashon which results in so called incurable diseases. The interaction of the positive Chetnon can set right the defects of the Akashons. The positive Chetnon penetrate the human body in great abundance during meditation. The human body acts like a lens, which concentrates both the positive and negative Chetnons, but during meditation, the flow of negative Chetnon is restricted. The great saints are source of positive Chetnons and these can be passed on to other persons also. A sick woman was suddenly healed when she touched Jesus' in the crowd and Jesus said, "Someone touched me, for I perceive that power has gone forth from me".

The evolution is continuous process. All that is manifested is subject to change, therefore, the principle of inverse variation is not permanent in nature. The so-called constants involved in the principle of inverse variation are changing very very slowly on Brahmanical scale of time. Nothing that is manifested is eternal but the Showman and show are eternal as it is expressed in the Shvetaashvatara Upanishad.

Epilogue

"For He these living worlds didst make,
  the flitting forms that rise and fade
upon the endless screen of life,
  behind which shines the wavering light.
Of the three gunas and when they cease
  Creation back to him doth sweep,
For time may pass and worlds may change
  but He is ever one, the same."

www.ingramcontent.com/pod-product-compliance
Lightning Source LLC
Chambersburg PA
CBHW081715170526
45167CB00009B/3583